MACROECOLOGY

MACROECOLOGY

JAMES H. BROWN

WITHDRAWN

 The University of Chicago Press • Chicago and London

James H. Brown is Regents' Professor of Biology at the University
of New Mexico. He is coeditor, with Leslie A. Real, of *Foundations
of Ecology: Classic Papers with Commentaries,* published by the
University of Chicago Press in association with the Ecological
Society of America.

The University of Chicago Press, Chicago 60637
The University of Chicago Press, Ltd., London
© 1995 by The University of Chicago
All rights reserved. Published 1995
Printed in the United States of America

04 03 02 01 00 99 98 97 96 95   1 2 3 4 5

ISBN 0-226-07614-8 (cloth)
    0-226-07615-6 (paper)

Library of Congress Cataloging-in-Publication Data

Brown, James H.
    Macroecology / James H. Brown.
        p.   cm.
    Includes bibliographical references and index.
    1. Ecology.   I. Title.
QH541.B75   1995
574.5—dc20   94-31250
                                                  CIP

♾ The paper used in this publication meets the minimum requirements of the American National Standard for Information Sciences—Permanence of Paper for Printed Library Materials,
ANSI Z39.48–1984.

for Astrid

Contents

Preface

In the original Spanish edition of his stimulating and insightful book *Aerography*, my friend Eddie Rapoport (1982) bared his soul in the preface. I do the same here and in the remainder of this book.

Most scientists seem to have a single philosophy and a well-defined, unitary approach to their work. Most of my colleagues spend their entire careers studying a particular level of biological organization, a single taxonomic group of organism, and one type of habitat. Furthermore, they tend to be either experimental, hypotheticodeductive reductionists or nonexperimental, inductive holists.

My colleagues who call themselves ecologists are virtually all reductionists. They use a hypotheticodeductive, usually experimental approach to take apart small, well-defined ecological systems and try to understand how they work. This is true of the vast majority of ecologists, regardless of whether they are physiological and behavioral ecologists who work at the level of the individual, or population and community ecologists who study the structure and dynamics of assemblages of many individuals and often multiple species, or ecosystem ecologists who investigate the transfer of energy and nutrients in aquatic or terrestrial habitats.

On the other hand, my colleagues who call themselves biogeographers, paleobiologists, and macroevolutionists tend to be holists. They use inductive, nonmanipulative methods to study whole systems or emergent characteristics of large, complicated assemblages of many species distributed over geographic spatial scales and evolutionary time scales. But they, too, want to understand how these systems work.

I am an oddball. I have always been fascinated by the variety of living things. In my efforts to understand this diversity, I have found it impossible to confine my studies to a single level of organization, kind of organism, or type of habitat. I have also been unable to be either an experimental, hypotheticodeductive reductionist or a nonexperimental, inductive holist. It has always seemed to me that each approach can answer questions that the other cannot, so I have tried to practice both. I have two research programs. In one, I use controlled field experiments to dissect the system in an effort to understand the relationships among the species and functional groups of both animals and plants that coexist on 20 hectares of the Chihuahuan Desert. In the other, I collect and analyze nonexperimental data in an effort to understand the variation in the abundance, distribution, and diversity of birds and mammals across continental landscapes and over long periods of time.

This monograph is about my second, nonexperimental research program—the one I call macroecology. It is the one that the majority of my colleagues, most of whom consider themselves ecologists, seem to find less interesting. This view is reflected in the fact that my macroecological research has been less reliably funded and less frequently cited than my experimental studies. Yet I have persevered with the macroecological research because I continue to find its results just as interesting as the outcomes of the experiments. While lacking some of the rigor and power of the experimental approach, the macroecological research offers greater potential for generality and synthesis. Furthermore, the methodology, data, and ideas of macroecology have potentially important implications for other disciplines, such as biogeography, paleobiology, systematics, and conservation biology, in which experimental manipulation is usually impossible. This monograph is an effort to summarize, synthesize, and apply that part of my research program that I call macroecology.

On the one hand, this is a very personal book. I want to present my own vision of what macroecology is and what it might become. Macroecological research has profoundly changed the way I think about the world, and I want to share some of that sense of discovery and excitement. I know that much of what I say here will be controversial. Some of the data will be suspect, and some of the ideas will turn out to be incomplete or downright wrong. I take full responsibility for these failings. The process of science will insure that serious errors of fact or interpretation will be corrected. I can only hope that the shortcomings will be outweighed by the positive contributions.

On the other hand, I am acutely aware how much other individuals have contributed to the ideas and data in this book. My macroecological research program could never have been developed and this monograph could never have been written without their help. I am especially indebted (1) to earlier investigators, who paved the way by asking similar questions, collecting many of the data, and anticipating many of the ideas; (2) to contemporary colleagues, who have generously shared their data and ideas (including unpublished data and manuscripts); (3) to present and former students and postdocs, who have helped to assemble the data, interpret the patterns, and test the hypotheses; and (4) to friends and family, who have shared the excitement during the good times and encouraged me to persevere during the tough times. I will not attempt to identify all of these individuals, because the list would be too long. Some of these people are dead. The rest of you know who you are. I am grateful to all.

The first draft of this book was written while my wife and I were on sabbatical leave in Australia. A fellowship from the John Simon Guggenheim Memorial Foundation provided financial support. Barry and Marilyn Fox and Steve and Faye Morton helped in countless ways to make our stay in their wonderful country both productive and enjoyable. The CSIRO Centre for Arid Zone Research in Alice Springs provided housing and a conducive place to think and work. After returning from Australia, I was able to finish the manuscript with support from the University of New Mexico's Faculty Scholars Program, and with the assistance of Pablo Marquet, Sarah Linehan, Shahroukh Mistry, Qinfeng Guo, and Doug Kelt. Throughout the development of my ideas on macroecology, I have worked closely and had many discussions with Astrid Kodric-Brown, Mark Lomolino, Pablo Marquet, Brian Maurer, Dave Mehlman, and George Stevens. I am especially grateful to them, to Bob Holt and Larry Heaney, and to an anonymous reviewer for reading the entire manuscript and making many helpful comments and suggestions. As always, it has been a pleasure to work with Susan Abrams at the University of Chicago Press.

1 Introduction

LET'S BEGIN WITH AN EXAMPLE

A Practical Problem: Predicting Extinctions Caused by Global Climate Change

There is presently widespread concern that the earth's climate will warm by at least 3°C within the next century. Such environmental change, comparable to the shifts in climate that accompanied the advances and retreats of glaciers during the Pleistocene, would have drastic effects on the distribution and diversity of living things. But just what effects? Which species and taxonomic groups would be most affected? How much would their geographic ranges expand or contract? Which ones would go extinct?

Consider a region where I have done much fieldwork, the Great Basin of western North America (fig. 1.1). Here, rising out of the extensive sagebrush desert, are numerous isolated mountain ranges. Each is a biological reserve in the sense that the relatively unspoiled habitats of woodland, forest, meadow, and stream at its high elevations, isolated from other

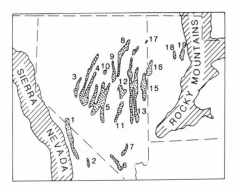

FIGURE 1.1. Map of the Great Basin region of western North America, showing the isolated mountain ranges. The numbers identify the ranges referred to in table 1.1 and figure 1.3.

1

mountaintops by the intervening desert valleys, preserve a special set of plant and animal species. Many of these boreal plants and animals are survivors of populations that were much more widely distributed more than 10,000 years ago, during the Pleistocene, when the climate was cooler and wetter and suitable habitats were broadly distributed across the Great Basin at lower elevations.

Clearly the anticipated global warming, if it occurs, will exacerbate the phenomenon that occurred at the end of the Pleistocene, forcing the cool, moist habitats to even higher elevations, causing them to shrink in size, and threatening species with extinction. Let's focus on just one group of particularly threatened species, the small terrestrial mammals. Many are relicts that presently survive on only a small fraction of the mountain ranges that they inhabited 10,000 years ago (Brown 1971a; Grayson 1987). With additional climatic warming of 3°C, which species are most likely to go extinct on which mountain ranges?

The Traditional Microscopic Approach

Faced with this question and given sufficient resources of money, equipment, and personnel, the typical ecologist or conservation biologist would head for the field to study the population biology and community relationships of selected populations of small mammals on particular mountain ranges. He or she would measure such things as distribution, abundance, habitat associations, and genetic diversity, and then attempt to predict how the anticipated climatic change would affect the populations by changing these conditions. The results of such a study, while valuable, would be limited. There would never be enough time and resources to study all species populations on all mountaintops. A great deal of questionable extrapolation to other populations of the same species on other mountain ranges and to different, unstudied species on the same mountain ranges would be required.

The Macroecological Approach

Faced with the same question and given one semester and virtually no resources, Kelly McDonald and I went not to the field but to the library (McDonald and Brown 1992). From information available in the literature, we were first able to predict how much of an elevational shift in habitat would be caused by the assumed 3°C climatic warming: the lower limit of woodland would shift approximately 500 m above its present elevation (2,280 m) (fig. 1.2). Next we used topographic maps to determine by planimetry the present and expected future areas of boreal habitat on each mountain range. Since the present occurrence of each small mammal species on each mountain range is reasonably well known (but see

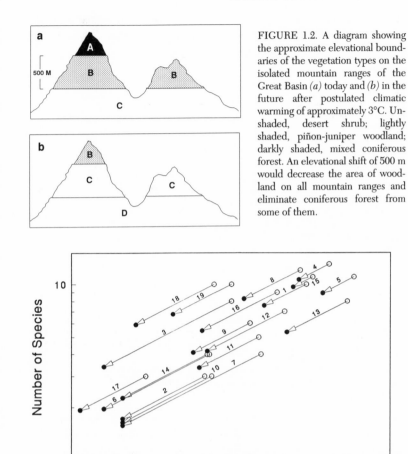

FIGURE 1.2. A diagram showing the approximate elevational boundaries of the vegetation types on the isolated mountain ranges of the Great Basin (*a*) today and (*b*) in the future after postulated climatic warming of approximately 3°C. Unshaded, desert shrub; lightly shaded, piñon-juniper woodland; darkly shaded, mixed coniferous forest. An elevational shift of 500 m would decrease the area of woodland on all mountain ranges and eliminate coniferous forest from some of them.

FIGURE 1.3. The species-area relationship, illustrating the distribution of boreal mammal species richness among the isolated mountain ranges of the Great Basin as a function of the area above 2,280 m elevation. The arrows show the changes in area and numbers of species predicted to be caused by climatic warming: the open circle at the base of each arrow indicates the present number of species, and the solid circle at the point of the arrow indicates the number predicted to remain after equilibration with a 3°C increase in average temperature. Numbers identifying the mountain ranges correspond to those in figure 1.1. (From McDonald and Brown 1992; reprinted by permission of the Society for Conservation Biology and Blackwell Scientific Publications, Inc.)

Grayson and Livingston 1993; Kodric-Brown and Brown 1993b), we were able to use the species-area relationship (fig. 1.3) to estimate the area of boreal habitat required to support a given number of species, and thus to predict the number of species that would be expected to go extinct on

TABLE 1.1. Distribution of fourteen small boreal mammal species among nineteen isolated mountain ranges in the Great Basin at present and after predicted extinctions due to effects of global warming.

Species	Mountain ranges[a]									
	4	8	5	1	15	19	16	13	3	18
Eutamias umbrinus	X	X	X	X	X	X	X	X	X	X
Neotoma cinerea	X	X	X	X	X	X	X	X	X	X
Eutamias dorsalis	X		X	X	X	X	X	X	X	X
Spermophilus lateralis	X	X	X	X	X		X	X	E	
Microtus longicaudus	X	X	X	X	X	X	X	X	E	X
Sylvilagus nuttalii	X	X	X	X	X		E	E	E	
Marmota flaviventris	X	X	X	X	X	X	E	E	E	X
Sorex vagrans	X	X	X	X	X	X	E	E		X
Sorex palustris	X	X	X	X	E	X				E
Mustella erminea	X			E	E	E	E			E
Ochotona princeps	X	E	E	E					E	
Zapus princeps	E	E	E			E				
Spermophilus beldingi	E	E								
Lepus townsendii		E				E				
Present number of species	13	12	11	11	10	10	9	8	8	8
Predicted number of species	11	8	9	10	8	7	5	5	3	6

Source: McDonald and Brown 1992. Reprinted by permission of the Society for Conservation Biology and Blackwell Scientific Publications, Inc.

[a]Numbers identifying the mountain ranges correspond to those in figure 1.1.

X, species now present and expected to persist; E, species now present and predicted to go extinct.

each mountain range when the habitat became reduced. Then, knowing the number of species predicted to go extinct, we were able to use the "nested subset" structure of the mountaintop mammal faunas (table 1.1)[1] to determine which particular species were most likely to be lost from each mountain range.

The answers are sobering. Different ones of the nineteen isolated mountain ranges are predicted to lose 9%—62% of their present boreal small mammal species. Of fourteen species, three are predicted to go extinct on all the mountaintops where they presently occur, two are predicted to survive on all mountain ranges, and the remainder are expected to disappear from some of their present range.

1. The list of species in table 1.1 has been augmented by the addition of a few records of additional species on a few mountain ranges (see Grayson and Livingston 1993; Kodric-Brown and Brown 1993b). I have not redone the analyses based on the updated list, because the results would be very similar. The point of this modeling exercise is not to predict exactly what will happen when the climate warms 3°C, but to help us to think in a precise, logical way about the potential effects of global warming on the biotas of isolated habitats and biological reserves.

									Number of ranges	
12	9	11	7	6	14	10	2	17	Present	Predicted
X	X	X	X	X	X	X			17	17
	X	X	X		X	X	X	X	17	17
X	X	X	E		E	E	X	X	17	14
X	X	X	E		E			E	14	10
E		E		X					13	10
E	E	E					E		12	5
	E								11	7
E									10	7
				E					8	5
									6	1
									5	1
				E					5	1
									2	0
									2	0
7	6	5	4	4	4	3	3	3		
4	4	3	2	2	2	2	2	2		

These predictions should be taken with caution. In order to make them, we had to make several important assumptions: about the magnitude of regional temperature change, its effect on elevational shifts in boreal habitat, the response of the small mammal populations to the reduced habitat areas, and so on. Traditional ecologists or conservation biologists, however, would have to make many of the same assumptions, and they would have to extrapolate from the small number of species populations and mountain ranges where their fieldwork was conducted. Given the limitations of time and resources, our predictions may well be at least as accurate as those that can be made from detailed field studies.

The point of this little exercise is that a macroscopic approach often offers an alternative way to address certain kinds of ecological and biogeographic questions. In this case it was a practical question of conservation biology, but one that can be approached by applying some basic concepts and data. Kelly McDonald and I were able to predict the likelihood of extinction of individual mammal species on particular mountaintops without measuring the details of population biology and community ecology. It is not that such factors as population sizes, habitat requirements, and genetic variation do not affect the probability of extinction—they clearly do. But the effects of these variables on extinction are reflected and integrated in the present distributions of species among mountaintops—so well reflected, in fact, that the macroscopic approach offers an alternative way to make predictions about the effects of global climate change on the montane mammals of the Great Basin.

The above example illustrates some of the features, and both the strengths and the weaknesses, of what I call the macroecological research program.

THE BACKGROUND OF MACROECOLOGY

Spectacular advances have been made in ecology, biogeography, systematics, paleontology, and evolutionary biology in just the last two or three decades. Much has been learned about the processes that regulate the abundance, distribution, and diversity of species in local habitats; the effects of history and contemporary environment on the geographic distributions of species; the phylogenetic relationships among organisms; and the fossil history of life on earth. Most of this progress is due to advances within traditional disciplines: to application of mathematical models and rigorous experimental methods in ecology, to new conceptual approaches to historical and ecological biogeography, to development of robust theoretical and molecular methods to reconstruct phylogenetic history, and to insights into the processes of diversification and extinction obtained from the fossil record.

But despite the progress within the specialized disciplines, many of the fundamental questions have remained unanswered and many new ones have been raised. It is becoming increasingly apparent that the answers to many of these questions lie beyond—and across—the boundaries of the traditional disciplines. Ecologists recognize the need to know the effects of phylogenetic constraints and large-scale geologic, climatic, and biogeographic processes in order to interpret the response of small-scale systems to natural variation and experimental perturbations. Biogeographers and systematists want to know how past environmental conditions influenced the distribution and diversification of lineages of related species. Paleontologists and macroevolutionists want to understand the causes of the varying rates of evolution and the pulses of speciation and extinction that they see in the fossil record. Addressing these questions will require research programs that cross the boundaries and explore the interfaces between the traditional disciplines.

This book describes one research program that attempts to address some of these interdisciplinary questions. Brian Maurer and I (Brown and Maurer 1989) have called it "macroecology." It is a nonexperimental, statistical investigation of the relationships between the dynamics and interactions of species populations that have typically been studied on small scales by ecologists and the processes of speciation, extinction, and expansion and contraction of ranges that have been investigated on much larger scales by biogeographers, paleontologists, and macroevolutionists. It is an effort to introduce simultaneously a geographic and a historical perspective in order to understand more completely the local abundance, distribution, and diversity of species, and to apply an ecological perspective in order to gain insights into the history and composition of regional and continental biotas.

THE GOALS OF THIS BOOK

This book presents my vision of the macroecological research program: what it is, its strengths and weaknesses, its accomplishments, the questions that remain unanswered, promising areas for future research, and some applications to practical problems of human ecology and conservation biology. I will not try to survey all of the relevant literature or to give "equal time" to other research programs that are tackling similar kinds of interdisciplinary problems. I have tried to keep up with some of the advances in other disciplines by talking to colleagues and selectively reading their papers, but I cannot claim to have mastered these diverse subjects.

This is a book of blatant advocacy. I will emphasize the positive features of the macroecological research program, the new insights that have come from its particular conceptual and methodological perspective, and the applications of the data and ideas to problems of interest to both basic and applied scientists. I also want to convey the optimism and excitement that I have experienced as I have pursued some of the topics. For me, an important mark of interesting science is that it changes my view of the natural world and challenges rather than confirms existing dogma. Macroecology has done this, repeatedly revealing patterns and processes that I did not expect on the basis of existing paradigms or my own biological intuition. It has offered many fresh new insights, including new ways of interpreting perplexing results of my small-scale, long-term experimental studies.

This book is not intended to be a complete or final statement about any of the topics that it explores. It is a progress report. The empirical patterns that are presented represent only a fraction of the data that could be collected and of the analyses that could be performed. The processes that are invoked should be regarded as hypotheses that still need to be tested thoroughly. There is much scope for rigorous mathematical modeling to strengthen the conceptual basis of macroecology.

The ultimate value of the book will be determined by its impact on you, its readers. If the ideas are interesting, the approach seems useful, and the preliminary data are intriguing, then I encourage you to explore them further. This book will have achieved its goal if it contributes significantly to its own obsolescence—if it stimulates additional macroecological research and the writing of more definitive papers and books.

THE PLAN OF THIS BOOK

What follows, then, is my progress report. I have organized it into thirteen chapters. This first chapter explains my motivation for writing the book and now tells the reader what to expect. The two following chapters describe the domain of macroecology. Chapter 2 provides a definition of macroecology. It presents the rationale for a macroscopic, statistical, nonmanipulative approach that can overcome some of the inherent limitations of traditional "microscopic" experimental ecology and provide new insights into the structure and dynamics of complex ecological systems. Chapter 3 develops the concepts of species, niche, and community that will be used throughout the book.

The next three chapters form a subsection that might be entitled "Theory and Data." They present the topics that I have investigated most thoroughly. These chapters are largely inductive in that they draw together empirical patterns and search for logical connections among them. Chapter 4 examines the relationship between the local abundance and the geographic distribution of species. Chapter 5 adopts the perspective of allometry to investigate the effects of body size on the abundance, distribution, and diversity of species. Chapter 6 is concerned with the geographic ranges of species: their sizes, shapes, and boundaries.

The theories induced to explain the empirical patterns can be cast in a more deductive framework because they invoke specific mechanisms that have additional logical consequences. The next section, which could be entitled "Mechanisms and Predictions," considers the relationships between macroecological patterns and mechanistic processes at three levels: the structure and function of individual organisms (chap. 7), the structure and dynamics of populations (chap. 8), and the origination and extinction of species (chap. 9). A general message of these three chapters is that the macroecological approach can offer new insights into traditional subjects, such as physiological adaptations, population dynamics, community organization, geographic range shifts, and evolutionary diversification.

Some of these topics are pursued further in the last major section, which might be called "Explorations and Applications."Chapter 10 considers how, by providing a common energetic and thermodynamic perspective on levels of organization from individuals to ecosystems, macroecological patterns and processes might contribute to a unification of the subdisciplines of traditional ecology. Chapter 11 explores the implications of increasing the spatial and temporal scale of ecology so as to break down the distinction between "history" and "ecology" in biogeography and macroevolution. Chapter 12 applies data and theory of macroecology to conservation issues by focusing on the unprecedented ecological dominance of our own species and its concomitant impacts on the global environment and other organisms.

Finally, in chapter 13, "Reflections and Prospects," I conclude the book by singling out a few of its most novel and important contributions. I also speculate on the future of the macroecological research program.

CONCLUDING REMARKS

This book develops a particular methodological, empirical, and theoretical approach that I call macroecology. From this perspective I explore the interface of the traditional disciplines of ecology, biogeography, and macroevolution. I hope to offer new insights into some basic questions about the abundance, distribution, and diversity of living things.

2 The Macroecological Approach

WHAT IS MACROECOLOGY?

Macroecology is a way of studying relationships between organisms and their environment that involves characterizing and explaining statistical patterns of abundance, distribution, and diversity. In describing the approach and domain of macroecology, Brian Maurer and I (Brown and Maurer 1989) used the prefix "macro" for two reasons. First, in order to characterize patterns in the statistical distributions of variables among individuals, populations, or species, it is usually necessary to work at relatively large spatial and/or temporal scales so as to obtain sufficiently large samples. Second, in order to investigate the implications of important advances in other disciplines such as biogeography, paleobiology, systematics, and the earth sciences, it is necessary to expand the scale of ecological research.

I view macroecology not as a new and separate field of science, but as a research program in ecology with some distinctive characteristics. Macroecology differs from most of recent and current ecology in its emphasis on statistical pattern analysis rather than experimental manipulation. This difference is more a matter of practicality than of philosophy, however. It is often impractical, impossible, or immoral to perform replicated, controlled experiments on the spatial and temporal scales required to answer many basic and applied questions. Consequently, it is necessary to find alternative ways to make inferences about the natural world (MacArthur 1972b; Orians 1980). Because of its emphasis on extant statistical patterns rather than on responses to artificial perturbation, macroecology may at first appear to be more inductive than most ecology. I will show, however, that macroecological research involves not only the inductive process of developing hypotheses to explain observations, but also the deductive process of making additional observations to test the hypotheses.

Macroecology is self-consciously expansive and synthetic. In this respect it does differ philosophically from much of traditional ecology, which I would characterize as becoming increasingly reductionist and specialized.[1] Rather than trying to use ever more powerful microscopes to study the fine details of ecological phenomena, macroecology tries to develop more powerful macroscopes that will reveal emergent patterns and processes. To make an analogy, the goal is not to understand a tapestry in terms of warp and woof and the chemistry of fibers and dyes, but to see and interpret the entire scene. In order to visualize the big picture it is necessary to stand back and take a distant view. Accordingly, macroecology attempts to increase the spatial and temporal scale of ecological inquiry, and also to expand the kinds of questions asked and the range of phenomena studied. It tries to achieve synthesis by exploring the relationships between ecological phenomena and the patterns and processes studied by basic and applied scientists in other disciplines.

THE CHALLENGE OF COMPLEXITY

Background

There seems to be a predilection among scientists, at least among biologists, to take a microscopic, reductionist approach to research. We are impressed by the triumphs of particle physics and molecular biology, and perhaps equally impressed by the difficulties and less rapid progress of cosmology and ecology (see recent critiques by Simberloff 1980; McIntosh 1985; Peters 1991). This experience suggests that it is profitable to take complex systems apart, identify their components, and figure out how they work. It is much more difficult to understand the configurations and dynamics of entire, intact complex systems, whether these be macromolecules, individual organisms, or assemblages of many species. It is not easy, perhaps not even possible, to recreate the properties of the complex systems from a knowledge of their components and the interactions among them.

In ecology and evolutionary biology this prejudice in favor of reductionism has been reinforced by the recent successes of experimental ecological and molecular evolutionary studies. Many important results have

1. An extreme example is the subdiscipline of chemical ecology, which has taken an exceptionally reductionist, high-technology approach to studying chemically mediated processes, such as plant defense against herbivores. Many compounds important in ecological interactions have been isolated and their chemical formulae and structures characterized. In order to obtain this level of reductionist precision, however, much of the research is being done by chemists who have little knowledge of or interest in the effects of these compounds on free-living individuals, populations, communities, and ecosystems.

come from research programs that have focused on microscopic components and their interactions, whether these be individual organisms within local ecosystems or nucleotides within genomes. Controlled, replicated experimental manipulations have enabled ecologists to identify processes that regulate the dynamics of populations and the organization of communities. Applications of molecular techniques have enabled systematists to reconstruct phylogenetic relationships and evolutionary biologists to better understand the genetic basis of evolutionary change. No one should discount the importance of these advances.

It would be equally wrong, however, to claim that these studies have answered all of the interesting questions in their disciplines. The results of microscopic studies cannot simply be extrapolated to larger scales to explain macroscopic patterns and processes. In fact, the more that we learn about the microscopic organization of complex systems, the more obvious is the need for macroscopic studies to place the findings in broader perspective. Ecological experiments and other small-scale studies have revealed a great deal about the influence of abiotic conditions and biotic interactions on the structure and dynamics of local populations, communities, and ecosystems. However, unless experiments are replicated in several different habitats and localities—and there seldom are sufficient resources to do so—it is impossible to know which results are specific to the particular system and which can be generalized to other systems. More importantly, all natural ecosystems are open to the exchange of energy, materials, and organisms across their necessarily arbitrarily defined boundaries. Consequently, large-scale climatic, oceanographic, and geologic factors, which are difficult or impossible to manipulate experimentally, affect biogeochemical processes, population dynamics, and community structure within local ecosystems (e.g., Roughgarden, Gaines, and Pacala 1987; Roughgarden, Gaines, and Possingham 1988; Molles and Dahm 1990). Large-scale events of speciation, colonization, and extinction also profoundly influence the biotic composition of local ecosystems (e.g., Ricklefs 1987; Pulliam 1988; Brooks and McLennan 1991; Ricklefs and Schluter 1993). Because most of the processes that operate at larger spatial scales have long-lasting effects, their influence often appears to be a consequence of "history."

Macroscopic studies are also needed to place the contributions of molecular biology to systematics and evolutionary biology in a broader perspective. Molecular techniques now permit reconstruction of phylogenies, determination of breeding systems, and analyses of genome organization. Important as these results are, however, their value would be greatly enhanced if they could be placed in an environmental and historical context. To what extent have environmental conditions caused or contributed to speciation, anagenetic evolutionary change, geographic range shifts, and extinction? To what extent are the radiations of lineages truly "adaptive," in the sense that they reflect the differential survival and proliferation of individuals or species with traits that confer specific advantages in particular environments? To what extent does the organization of genetic material within and among chromosomes or between its nuclear and cytoplasmic constituents reflect phylogenetic constraints and/ or adaptive responses to particular environments? Answers to these questions require placing the processes of molecular evolution in the ecological, biogeographic, and paleontological contexts in which they occur— and in which they occurred in the past.

Finally, there is a desperate practical need for macroscopic studies. Modern humans are changing the world in ways that are unprecedented in the history of life on earth and that threaten the survival of vast numbers of species, including our own. The most severe environmental problems caused by humans are on regional to global scales. These problems cannot be addressed solely by traditional, small-scale ecological experiments, in part because there is not sufficient time, money, and personnel to do reductionist studies of each habitat, each species, and each process. Even if such small-scale, experimental studies were done, however, many of their results could not validly be extrapolated to regional and global scales. To address regional and global problems of environmental change and decreasing biological diversity will require macroscopic studies that necessarily trade off the precision of small-scale experimental science to seek robust solutions to big problems.

Complex Adaptive Systems

Ecologists and evolutionary biologists have set themselves the task of understanding the diversity of life—of characterizing the variety of living things and discovering how such variety was produced and how it is maintained. There is no escaping the magnitude of this endeavor.

Consider the diversity of contemporary organisms. No one knows how many species of organisms inhabit the earth, but the "best estimates" range between 5 million and 50 million (e.g., Erwin 1983a; Wilson 1988; May 1988). Similarly, no one knows how many species occur together within a local ecosystem, but the number probably ranges from a handful in such extreme environments as hot springs, hypersaline lakes, and antarctic deserts, to perhaps tens of thousands in tropical forests (e.g., Erwin 1983b). These living organisms are only a tiny fraction of the number of species that once inhabited the earth during its 5-billion-year history; more than 99.99% of the species that ever lived are now extinct (Nitecki 1984; Raup 1986). The variety of life is reflected not only in the number of species, but also in their attributes. Consider that organisms vary by more than twenty orders of magnitude in body size, from 10^{-13} g to 10^8 g. There is comparable variation in life span (from a few minutes to thousands of years) and in use of space (some lichens live for decades on a few square centimeters of rock, whereas some birds and whales migrate tens of thousands of kilometers every year). The size of genomes varies from about 10^4 nucleotides in the simplest viruses to more than 10^{10} nucleotides in some flowering plants and vertebrates.

The task of ecology and evolutionary biology is to derive the laws that govern this variety of life. The goal is to develop a body of theory and data that organizes what we already know about organic diversity and that generates new information and ideas, new data and hypotheses, to guide further study. If this seems daunting, consider how far we have come in just the two centuries since the revolutionary insights of Lyell, Darwin, and Mendel, or in the few decades since the beginning of modern experimental ecology, molecular and phylogenetic systematics, and plate tectonics. There is reason for optimism, but much remains to be done.

Ecologists and evolutionary biologists can take some comfort in the fact that they are not alone in their struggle to deal with such daunting complexity. There is an emerging science that seeks to identify the common features of complex adaptive systems and to develop conceptual approaches and technological tools for studying them (e.g., Lewin 1992; Waldrup 1992; Kauffman 1993; Cowan, Pines, and Melzer 1994). Complex adaptive systems have several common features: (1) they are composed of numerous components of many different kinds, (2) the components interact nonlinearly and on different temporal and spatial scales, (3) the systems organize themselves to produce complex structures and behaviors, (4) the systems maintain thermodynamically unlikely states by the exchange of energy and materials across their differentially permeable boundaries, (5) some form of heritable information allows the systems to respond adaptively to environmental change, and (6) because the direction and magnitude of any change is affected by preexisting conditions, the structure and dynamics of these systems are effectively irreversible, and there is always a legacy of history. Examples of such complex systems include human brains, languages, cities, and certain kinds of computer programs.

Individual organisms can be regarded as a special class of complex adaptive systems because they exhibit all of the above characteristics. Elsewhere (Brown 1994, in press) and in chapter 3 I suggest that species also possess most of these attributes. I do not want to argue here whether populations and communities are most productively viewed as complex adaptive systems or as assemblages composed of complex adaptive systems. Populations and communities also exhibit many of the above properties, but they differ from individuals and species in at least one important respect: they do not have clearly defined, selectively permeable boundaries. Individual organisms have well-defined beginnings (births) and ends (deaths), insides (bodies) and outsides (environments). I will claim that species also exhibit a similar discreteness. Populations and communities are different. The absence of discrete boundaries and the unregulated movement of organisms, as well as of energy and nonliving materials, makes the identification of most populations and communities quite arbitrary. A population or community is simply the convenient assemblage of organisms that a scientist selects for study. Usually it is defined simply in terms of the organisms that occur together at some arbitrary study site for some arbitrary period of time.

The Nature of Ecological Complexity

Every child who has toyed with an old-fashioned jeweled-movement watch knows that it is far easier to take a complex system apart than it is to reassemble the parts and restore the important functions. Similarly, no science has succeeded in understanding the structure and dynamics of a complex system from a reductionist approach alone. Physicists have not solved the three-body problem,[2] molecular biologists have not recreated an organism from its chemical constituents in a test tube, and neurobiologists have not been able to explain memory and cognition in terms of interactions among neurons. In ecology, we know, at least in principle, how to determine the number of species that occur together in a woodlot or pond, and how to characterize their trophic and phylogenetic relationships to one another. We do not know, even in principle, how to predict the changes in the number, identity, trophic relationships, and genetic constitution of the species that will occur if the average temperature increases 3°C or if an exotic species colonizes.

Mathematical and computer-based studies of complex systems show why this is true. A system comprising a modest number of different kinds of components interacting in specified ways will, over time, exhibit complex nonlinear dynamics and continually changing structures. Although the rules governing the organization of the system are known—because the investigator wrote the equations or the computer program—the specific outcomes are essentially unpredictable. Small differences in initial conditions, stochastic events, time lags, processes operating on different

2. The three-body problem in physics is an example. It is possible to use the laws of gravitation and motion to predict the dynamics of a two-body system, given the masses, velocities, and initial positions of each object. No one has yet managed to predict the behavior of the analogous three-body system.

time scales, and spatial subdivision all contribute to the complexity of organization and the variability of behavior. This is the bad news: even when the components and the rules for assembling them are known, it is essentially impossible to predict the details of the resulting complex system.

But here is the good news: at another, more macroscopic level the structure and behavior of complex systems are predictable. The dynamic organization of the entire system is constrained by the nature of its components and the kinds of interactions among them. These constraints produce emergent, predictable patterns of macroscopic, whole-system structure and dynamics. Some of these are revealed as statistical patterns. Thus, for example, while each run of a set of equations or a computer program may give a somewhat different outcome, many replicate simulations will produce a limited range of outcomes that can be characterized statistically as a probability distribution.

The Study of Ecological Complexity

In my high school physics class the teacher performed the following demonstration. He placed a vial of peppermint oil on his desk at the front of the room, removed the stopper, and asked the students to raise their hands when they first detected the scent of peppermint. The students raised their hands in a wave that began in the front of the room and progressed to the rear. We know why this pattern occurred. "Random diffusion" caused a net movement of peppermint molecules from the high concentration in the vial to the areas of lower concentration in the room. Each student raised a hand when a molecule of peppermint reached his or her olfactory epithelium and triggered a sensory nerve impulse.

Consider how physics has explained the behavior of a gas. Each molecule took a specific path from the vial to a student's nose. This path was marked by numerous changes in velocity and direction as the peppermint molecule moved among and collided with the other molecules in the room. A physicist would not try to describe and explain this path. Even if it could be done, such an exercise would require numerous measurements and produce little generality—it would be necessary to repeat the entire process to understand the path of anther molecule to another student's nose. Physicists have studied the movement of gas molecules, however. They have investigated the statistical behavior of many molecules, called the process diffusion, and developed mathematical equations to characterize it. Diffusion is a very general process: it is an emergent property of any large collection of gas molecules that seemingly move and collide at random. It explains the highly predictable pattern of detection of peppermint.

There is a lesson here for ecology. During the last few decades population ecologists have tried to describe and predict the fluctuations in the local abundance of species; community ecologists have tried to describe and predict changes in the composition of locally coexisting species. I would liken this to a physicist trying to map and predict the paths of single gas molecules. It is little wonder that progress has been slow and generality has been limited. It is important to realize that this is not because the causes of past changes in abundance or species composition are inherently unknowable or because prediction of the future trajectory is impossible. Because humans are more similar in size to other organisms than to gas molecules, it is probably easier for us to find out how small differences in environmental conditions, time lags, and other factors can affect the structure and dynamics of a population or community. But the generality and predictability will still be limited, because even very small differences can be amplified by nonlinear processes to produce divergent outcomes.

This is not to say that what I call microscopic studies of the structure and dynamics of populations and communities have little value. Such studies are important for two reasons. First, there are practical reasons for trying to describe and predict the behavior of certain systems, even though the generality may be limited. There are powerful socioeconomic reasons for understanding populations and communities of resource and pest species so that we can manage them for human benefit. There are aesthetic and moral reasons for understanding the population dynamics of endangered species and the organization of threatened communities so that we can take actions to conserve these systems.

Second, it is important to study the microscopic details of complex ecological systems because the nature of the components and their interactions ultimately determines the emergent statistical properties of these systems. To return to the analogy with the physics of gases, it was essential to know that gases are collections of seemingly randomly moving molecules and to understand the factors that affect their velocity and collisions in order to understand the process of diffusion. Similarly, any relatively complete understanding of population dynamics or community organization will require a knowledge of both the microscopic components, the emergent macroscopic properties, and the relationships between the two.

For these reasons, I do not intend to be critical of microscopic ecological studies or of the reductionist, experimental approaches that enable them to be done practically and rigorously. Half of my own research is devoted to such a study. I do want to stress the importance of realizing the limitations of such studies, the reasons for their limited generality and predictability, and the contributions that can be made by macroscopic studies of the emergent properties of complex ecological systems. I advocate macroecology and other macroscopic approaches[3] not as an alternative to traditional experimental population and community ecology, but as a complement. I believe that ecology will advance most rapidly when there is a healthy balance between microscopic and macroscopic approaches.

THE MACROECOLOGICAL RESEARCH PROGRAM

Characteristics

Much of what is said above may seem abstract and wide-ranging. It is time to be more concrete, to focus on the conceptual and operational tenets of the macroecological research program.

Let's start by expanding on the definition given at the beginning of this chapter. Ecology is the study of relationships between organisms and their environment. The subdisciplines of population and community ecology are concerned with questions of how the abundance, distribution, and diversity of species are affected by interactions with other organisms and with the physical or abiotic environment. Macroecology is one approach to answering these questions. Several important elements characterize the macroecological research program (see also Brown and Maurer 1989):

1. Macroecology is concerned with the statistical distributions of variables among large numbers of comparable ecological "particles." Usually these particles are either many individual organisms within species populations or many species within local, regional, or continental biotas. The individuals and species are not exactly identical; they vary in their characteristics. Macroecology seeks to discover, describe, and explain the patterns of variation. Much of the emphasis is on the shapes and boundaries of statistical distributions, because these appear to reflect intrinsic, evolutionary or extrinsic, environmental constraints on the variation. In order to characterize and compare these distributions, it is desirable—but not always possible—to have samples of hundreds or thousands of particles.

2. The variables of macroecological study are ecologically relevant characteristics of organisms. The kinds of attributes that can be used are necessarily limited by the requirement for samples of large numbers of individuals, populations, or species. Most of my research has focused on variables, such as body mass, population density, and area of geographic range, that

---

3. Some kinds of ecosystem ecology, which study whole-system patterns and processes of energy and material exchange, are examples of another approach that focuses on emergent properties of complex ecological systems. Studies of emergent properties of food webs are yet another example.

affect the use of space and nutritional resources. Body mass is correlated with the energetic, nutrient, and space requirements of individual organisms. Local population density indicates the number of individuals that coexist in and are supported by a small area. The size and configuration of the geographic range shows the area of space and range of environmental conditions within which all populations of a species occur. Note that each of the above variables characterizes a different level of organization: individual, population, and species, respectively. Note also that this is by no means an exhaustive list; these are just examples of the variables that I have most frequently used in my macroecological studies.

3. The assemblages of organisms used in macroecological studies may be defined on the basis of either taxonomic or ecological similarity. Most hypotheses and analyses require that the organisms being considered be not precisely identical but at least comparable, in the sense that they are subject to similar evolutionary and/or functional constraints. This requirement is met if we assume that taxonomic classifications reflect phylogenetic relationships and that ecological classifications (e.g., into guilds or life forms) reflect functional similarities. I am aware that neither of these assumptions is always correct, but existing hierarchical classifications of biological diversity provide a reasonable starting point. This book is based largely on insights from data on North American terrestrial mammals (i.e., excluding flying and aquatic species) and land birds (excluding those groups that are primarily associated with and obtain their food from freshwater and marine habitats). These groups were selected because I am familiar with them and because they have yielded sufficient standardized data of relatively high quality. Other kinds of organisms could equally well be studied, and in some cases other investigators are doing so.

4. Macroecology tends to focus on phenomena at regional to global spatial scales and decadal to millennial temporal scales. Again, this is to a large extent a practical limitation, imposed by the need for large samples. But it means that macroecology is often concerned with patterns and processes at much larger scales than the small study plots and short field seasons of most experimental ecologists. It means that macroecological studies must often consider regional and global environmental variation, earth history, species dynamics (speciation, extinction, and geographic range shifts), and phylogenetic relationships. It means that macroecology explores the domain where ecology, biogeography, paleobiology, and macroevolution come together, and thus has the potential to forge synthetic links among these disciplines.

5. The macroecological research program is both empirical and theoretical, both inductive and deductive. It is concerned with the relationship between pattern and process. It is based on the assumption that some of the general processes that regulate the abundance, distribution, and diversity of organisms are reflected in emergent patterns in the statistical distributions of individuals, populations, and species. Macroecological research seeks to discover and describe these patterns, and to develop and test hypotheses to account for them. While much of the initial inspiration

comes inductively, from the discovery of patterns in data, the validity of the ideas ultimately must be evaluated deductively, by casting them as hypotheses that make testable predictions. This book will show that macroecology has not only discovered some intriguing patterns, but has also begun to develop and test mechanistic hypotheses.

Perspectives

Macroecology differs from most recent and current ecology in its focus on larger spatial and temporal scales and in its substitution of a comparative statistical methodology for an experimental manipulative one. Macroecology is based on the premise that one way to understand the structure and dynamics of complex ecological systems is to discover and explain their emergent, whole-system properties—and one way to do this is to stand back and take a sufficiently "distant view" that the idiosyncratic details disappear and only the big, important features remain. Thus macroecology deliberately sacrifices a great deal of detail in order to try to see the big picture.

Earlier I made the analogy between the physics of collections of many gas molecules and the ecology of assemblages of many individuals or species. This is an instructive analogy. It shows the difficulties inherent in trying to understand the details of the behavior of an individual particle, whether this be the path of a single gas molecule, the movements of a single individual, or the fluctuations of a single local population. It also suggests the kinds of insights that can come from taking a sufficiently distant view that these details can be ignored but the emergent statistical properties of the whole system become more apparent.

Like all analogies, however, this one breaks down. The statistical physics of gases is based on collections of more than 10^{10} molecules, so that small differences among individual molecules can often either be ignored or treated as statistical averages. In contrast, macroecologists are fortunate if they can obtain data on a few hundred individuals, populations, or species. In such assemblages, the differences among the individual "particles" do matter. In fact, much macroecological research endeavors to characterize and explain the pattern of variation among the particles—i.e., variation in characteristics such as body size, local abundance, and area and shape of geographic range.

The recent successes of microscopic, experimental ecology have lead some zealots to claim that this is the only way to do sound ecological science. This view is stifling, misguided, and wrong. Well-designed, controlled, replicated, manipulative experiments are a proven way to good, rigorous science—but they are not the only way. I am afraid that most ecologists have been conditioned by their reductionist training in biology and overly impressed by the spectacular successes of their cellular and molecular colleagues—and perhaps they feel inferior because of their own more modest accomplishments. This is unwarranted.

In its effort to understand complex structures and dynamics, ecology has more in common with astronomy and the geologic sciences than with most of traditional biology, chemistry, and physics. Robert MacArthur once said to me, "Astronomy was a respected, rigorous science long before ecology was, but Copernicus and Galileo never moved a star." Similarly, geologists have recently developed an elegant theory of plate tectonics that accounts for many characteristics of the structure and dynamics of the earth's crust, and they accomplished this feat of modern science without experimentally manipulating the crustal plates or the fluid mantle below.

There is another justification for increasing the scale of ecological research and addressing questions that cannot be answered by the kind of experimental ecology that has predominated for the last few decades. This is the practical need to understand the effects of the human population and its technology, and to apply this knowledge to reduce environmental degradation, to slow the loss of biological diversity, and to approach a sustainable balance between the demands of the human population and the renewal of the earth's resources. Many of the most severe problems, such as changes in climate and land use, destruction of habitats, and extinction of species, are occurring at regional to global scales. These problems cannot be addressed simply by extrapolating from the results of microscopic, experimental studies because qualitatively different processes may assume importance at the larger scales. To address these applied problems, it is necessary to "scale up" ecological research. The macroecological approach offers one way to do this.

OTHER MACROECOLOGICAL RESEARCH

I hope that colleagues will not be offended by my use of the term "macroecology." I try to avoid technical jargon and the coining of new terms. But I need some convenient way to refer to my macroscopic, nonexperimental research program. I am also very much aware, however, of how much I have been influenced by related research, both past and present.

Historical Precedents

Charles Darwin, Alfred Russel Wallace, and other nineteenth-century naturalists could legitimately claim to be the first macroecologists. Their writings are filled with descriptions of large-scale patterns and hypothesized mechanistic explanations.

By the 1920s, three lines of investigation that laid a foundation for macroecology were well under way. John Willis (1922) plotted frequency distributions for the areas of geographic ranges of species in different taxonomic groups. He found that almost invariably these exhibited the same form, which he termed a "hollow curve": there were many species with small ranges but only a few with large ones (fig. 2.1). Willis speculated that the ranges of different sizes reflected the history of species formation, with recently derived species being confined to small areas near their origin and long-established species having spread to inhabit large areas.

Willis's study was perhaps the first empirical analysis of the quantitative, statistical distribution of an ecological or biogeographic variable among many species. It was followed by other studies that compiled and interpreted data on the frequency distributions of abundances (Fisher, Corbet, and Williams 1943; Preston 1948, 1962a,b; MacArthur 1957; Williams 1964) and body sizes (Hutchinson and MacArthur 1959; Van Valen 1973a). All of these early studies share one common feature: they discovered empirical patterns that have proven to be very general, but also very difficult to explain (e.g., see May 1975; Sugihara 1980; Harvey and Godfray 1987; Brown and Maurer 1989; Lawton 1990).

It was also in the 1920s that Alfred Lotka (1922, 1925) made his most

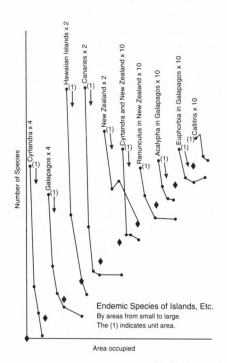

FIGURE 2.1. An illustration from Willis's interesting book *Age and Area*, showing the frequency distribution among species of vascular plants of the areas of their geographic ranges. Willis characterized these repeated patterns, in which the vast majority of species in each taxon and archipelago have very small ranges and only a few species have large ones, as "hollow curves." Because he used a statistical approach that focused on the distribution of attributes among species, Willis might be considered a pioneer in macroecology, although his interpretation of those patterns would be considered quaint today. (From Willis 1922.)

important contributions. Lotka brought to his ecological studies a background in physics. He used explicit analogies to statistical mechanics to develop mathematical models for the structure and dynamics of populations and communities. Although subsequent investigators have often misinterpreted the intent of Lotka's conceptualization, it provided an explicitly statistical approach to ecological and evolutionary theory. Robert MacArthur, during his short but enormously influential career from the late 1950s to the early 1970s, relied heavily on Lotka's pioneering work in his efforts to develop mathematical theories of ecology. While some of MacArthur's mathematical models are now considered too simplistic to capture realistically the workings of complex ecological systems (McIntosh 1985), others continue to provide valuable insights. Many of the latter take a macroscopic perspective and use species as the units of analysis (e.g., diversity of body sizes: Hutchinson and MacArthur 1959; colonization/extinction dynamics in biogeography: MacArthur and Wilson 1963, 1967; patterns and causes of geographic variation in species diversity: parts of MacArthur 1972b).

The third influential early scientist was Joseph Grinnell. His specific contribution is more difficult to pinpoint than Willis's or Lotka's, but he was a superb naturalist who saw important connections between what have become the specialized disciplines of ecology, biogeography, systematics, and evolutionary biology. Grinnell (1914, 1917; Grinnell and Swarth 1913) developed the concept of ecological niche, recognized the role of geographic barriers in allopatric speciation, and considered the processes of colonization and extinction in biogeography. All of these themes have received considerable attention since Grinnell's seminal contributions. Most notably, Hutchinson (1957) modified and formalized the concept of the niche, Mayr (1942, 1963) presented evidence for the process of speciation by geographic isolation, and MacArthur and Wilson (1963, 1967) developed a theory of insular biogeography based on an equilibrium between opposing rates of colonization and extinction.

Emphasis on the above three individuals and the very brief mention of subsequent, related contributions by other scientists cannot really do justice to the background already laid for macroecology prior to my own studies. Additional studies that laid the groundwork for various aspects of the macroecological research program will be mentioned throughout the book.

Ongoing Related Research Programs

It would also be misleading to suggest that the studies that my collaborators and I have done and are doing are the only ones that take a macroscopic approach to investigating the abundance, distribution, and diversity of species. Many of my colleagues recognize the potential of a statistical approach to ecological questions, do macroscopic research at the interface of ecology, biogeography, and evolution, and use methods similar to mine. Some of these individuals are independently doing research that is so similar to mine in conceptual basis and methodology that it could easily be called macroecology. Included in this category are studies of the distributions of body sizes, abundances, geographic ranges, reproductive characteristics, and other attributes among many species in different taxonomic groups (e.g., May 1978, 1986, 1988; Hengeveld and Haeck 1981, 1982; Hanski 1982c; Bock and Ricklefs 1983; Bock 1984a,b, 1987; Dial and Marzluff 1988, 1989; Gaston and Lawton 1988a,b, 1990a,b; Glazier 1988; Lawton 1990; Collins and Glenn 1990; Marzluff and Dial 1991). Perhaps the most encouraging thing about these independent studies with similar goals and methods is how frequently the same empirical patterns are found in different kinds of organisms or in different geographic regions. This suggests not only that the patterns are widespread, but also that the mechanisms that produce them may be very general.

Many more individuals are pursuing research programs that are distinctively different from mine. Although these focus on different questions or employ different methods, they have much to contribute to macroecology. Included in this category are phylogenetic approaches to ecological, biogeographic, and evolutionary questions (e.g., Brooks 1985; Gittelman 1985; Pagel and Harvey 1988, 1989; Futuyma and McCafferty 1990; Brooks and McLennan 1991; Harvey and Pagel 1991; Gittelman and Luh 1992; Taylor and Gotelli, in press); paleoecological and paleontological studies on changing patterns of species diversity over time and the effects of range shifts, speciation, and extinction (e.g., Bernabo and Webb 1977; Sepkoski 1978, 1982, 1984; Raup 1979, 1986; Martin and Klein 1984; Jablonski 1985, 1986a,b; Davis 1986; Graham 1986; Raup and Jablonski 1986; Delcourt and Delcourt 1987; Behrensmeyer et al. 1992); the new field of "landscape ecology" (e.g., Forman and Godron 1986; Turner 1987; Milne 1988, 1991, 1992; Danielson 1991; Pulliam and Danielson 1991; Johnson et al. 1992; Milne et al. 1992); and studies of the hierarchical organization of complex ecological and evolutionary systems (Allen and Starr 1982; Salthe 1985; O'Neill et al. 1987; Eldredge 1989). More will be said about the complementary relationships between macroecology and these other research programs throughout the book.

CONCLUDING REMARKS

The macroecological research program promises to do two things. First, by focusing on the emergent, statistical properties of complex ecological systems, it will contribute to answering some of the central questions of traditional ecology about the abundance, distribution, and diversity of species. During the last few decades ecology has developed from its roots in descriptive natural history into a rigorous, model-building, hypothesis-testing science. But the experimental approach that has dominated the empirical side of the discipline has severely limited both the questions that have been asked and the kinds of answers that have been obtained. A nonexperimental, macroscopic approach can identify patterns of abundance, distribution, and diversity on spatial and temporal scales much larger than those of traditional ecological studies. And it can give insights into the processes that produce these patterns that cannot be obtained from controlled manipulative experiments.

Second, macroecology promises to make synthetic connections: among different levels of organization, from individuals to populations to communities and biotas; among patterns and processes that occur on different spatial and temporal scales, from local to global and from diel to millennial; and among different scientific disciplines, such as ecology, biogeography, systematics, paleobiology, macroevolution, and the earth sciences. Such a synthesis will contribute to our understanding of how the living world is organized, how it came to be that way, and how it is now being changed by the activities of our own species.

3

Species, Niches, and Communities

Before going on to macroecology itself, it is necessary to develop some background. This chapter focuses on some fundamental properties of species, their relationships with their environments, and their associations with one another.

THE UNIQUENESS OF SPECIES

The Nature and Definition of Species

It is a tautology to say that each species is unique. Each species is described and recognized by the special features of its biology that set it apart from all other species, and especially from its closest relatives or sister species. Usually species can be identified on the basis of distinctive morphology, physiology, and behavior, but sometimes closely related species are very similar and molecular genetic traits must be used to distinguish them.

Despite some real problems in how to define a species in organisms as different as plants, animals, and microbes, species are convenient, natural units for ecological, biogeographic, and macroevolutionary analysis. Evolutionary biologists and systematists are seemingly always arguing the merits of alternative species concepts.[1] I will not become embroiled in this debate, except to make two points. First, I will argue that, in order for species to be operationally useful biological entities, they must be sufficiently distinctive to be (1) reliably recognized and identified, (2) useful units for evolutionary and paleontological studies, and (3) useful units for

---

1. Perspectives on species definitions run the gamut from Cracraft's (1989, 1990) concept of an evolutionary species as any population in which any unique derived trait is fixed in all individuals, to Williams's (1992) claim that all concepts of species as discrete, natural evolutionary and/or ecological units are fallacious.

ecological studies. Second, some kind of genetic, evolutionary, and ecological cohesiveness or boundedness is necessary if species are to be considered a class of complex adaptive systems comparable to individual organisms.

The vast majority of species meet the above criteria, especially within small scales of space and time. Consequently, an ecologist working in an area or with a taxon that is poorly known typically has little problem learning to recognize the species units, even though scientific names cannot be assigned to them until reference specimens have been identified by a taxonomist. The same is true for a paleobiologist working with a single sample of fossils. Sometimes there is a problem when the spatial or temporal scale of study is expanded. Then species may no longer be so discrete and easy to recognize; they may grade into one another over geographic space and evolutionary time (e.g., geographic races and chronospecies). This is a consequence of the fact that speciation, although it may be somewhat punctuated, takes time and is facilitated by spatial separation.

Species as Units of Macroecological Analysis

Since macroecology is concerned with large scales of both space and time, it must put up with this difficulty in determining the boundaries between closely related species. For the moment, at least, I am willing to be pragmatic and accept the current classifications of the taxonomic experts. All that is really necessary for my purposes is that species be operationally identifiable, that they represent relatively comparable units of biological organization, and that the individuals and populations recognized as being conspecific be more closely related to each other than to other recognized species. The species of terrestrial vertebrates and many other well-studied groups meet these criteria. Furthermore, an advantage in working with samples of hundreds of species is that, if the patterns and processes are robust, they will not be much affected by small changes in species-level taxonomy.

Species are convenient and appropriate units of analysis for macroecological studies for several reasons. First, as will be developed in more detail shortly, the concept of ecological niche offers a conceptual and empirical basis for treating the species as a distinct, recognizable unit with unique ecological requirements that are reflected in its abundance and distribution over both space and time. Second, a practical constraint of macroecological research is that it must be framed in terms of the kinds of data that are presently available or that might reasonably be obtained. Many of the appropriate data sets are either lists of species (e.g., those that occur in particular contemporary or paleontological samples) or quantitative characterizations of entire species (e.g., measurements of body sizes or other attributes of individual organisms, censuses of abundances, and maps of geographic ranges). Third, patterns of biological diversity are most dramatically reflected in the spatial and temporal variation in the composition of assemblages of species, and changes in species diversity reflect the dynamics of species origination and extinction processes. Finally, aside from ecology, the disciplines most relevant to the macroecological research program are biogeography and macroevolution. That species are also the primary units of study in both of these disciplines facilitates communication and synthesis.

In order to use species as units of macroecological analysis, I have usually characterized an entire species with a single value for each variable, ignoring intraspecific geographic or temporal variation.[2] At this preliminary stage of macroecological research, such simplification is justified. It should not create serious problems so long as intraspecific spatial and temporal variation is substantially less than the differences between species, which is nearly always true. Later, more refined macroecological studies may want to incorporate intraspecific variation.

The Place of Species in the Hierarchy of Biological Organization

We have all been taught that there is a hierarchy of biological organization that runs from molecules to ecosystems or the entire biosphere, and that includes cells, individuals, populations, and communities as important intermediate levels (table 3.1). While all these levels are useful subjects of study, I question the utility of this scheme as a hierarchy of structure and function. My alternative hierarchy would be similar from molecules to individuals and populations, but would then place species as the next, and final, level (table 3.1). I am not the first to propose such an alternative scheme (see Eldredge 1985, 1989).

The traditional hierarchy has three disadvantages. First, it mixes different kinds of units. All of the lower-level units are exclusively biological entities, but the ecosystem includes all components of the abiotic environment. This is inconsistent, because all of the lower-level units also exist in and are profoundly influenced by their own physical/chemical environments. Second, the two higher units in the traditional hierarchy, communities and ecosystems, are very difficult to characterize consistently (Allen and Starr 1982; O'Neill et al. 1987; Allen and Hoekstra 1992). They are defined in terms of arbitrary spatial and temporal boundaries. For ex-

---

2. This need not always be done. When appropriate data are available, it may be possible to use not only a single mean or median value, but also some measure of variation, such as standard deviation or range of extreme values. It may also be possible to include data on intraspecific geographic or temporal variation.

TABLE 3.1. Two representations of hierarchies of biological organization: the traditional hierarchy that appears in one version or another in most biology textbooks, and the ecological and evolutionary hierarchy of units of biological organization that I advocate (see also Eldredge 1985).

Traditional hierarchy	Ecological and evolutionary hierarchy
Molecules	Genes and enzymes
Subcellular organelles	Metabolic pathways
Cells	Cells
Tissues	Associated (colonies of) cells
Organs	
Organ systems	
Organisms	Organisms
Populations	Populations
Communities	Species
Ecosystems	Monophyletic lineages
Biosphere	Biosphere

ample, the stomach of a cow contains a diverse assemblage of microbes that plays a key role in the digestion of cellulose and other plant products. Thus we have a higher level, a community, actually living inside a lower level, an individual. Finally, and most importantly, communities and ecosystems have no inherent integrity or cohesiveness as biological entities.[3] They do not have discrete boundaries, share genetic information, reproduce, or evolve as a consequence of a Malthusian-Darwinian dynamic.

Using species as the next higher level above populations overcomes most of these difficulties. Species are consistent, equivalent units. They have relatively distinct boundaries in space and time, which include all of the equivalent lower-level units (i.e., populations or individuals). And they exhibit genetic cohesiveness and act as units of selection. I will return to this last point in chapter 11 because it is important for making connections between macroecology and macroevolution. The uniqueness of each species is reflected at all levels of organization, from the configuration of its molecules, to the morphology, physiology, and behavior of its individuals, to the abundance, distribution, and gene frequencies of its populations.

This hierarchical organization does not imply, however, that all of the

3. Wilson and Sober (1989) suggest that at least some kinds of communities can be viewed as superorganisms. There is some merit to this view. The eukaryote cells that make up all higher plants and animals evolved from a community of mutualistic species; the chloroplasts, mitochondria, and perhaps other organelles were once free-living microbes. The symbiotic origin of the eukaryotes was one of the most important events in the evolution of biological complexity: what presumably started out as an opportunistic Gleasonian community evolved into a discrete unit (first an individual and then, with the evolution of multicellular organization, a cell within an individual). Most ecological communities, however, do not act as discrete, functional units of ecology and evolution.

properties of the higher levels can be understood by reductionist studies at lower levels. Does a melanic mouse occur on volcanic soil because the individuals are dark-colored, or are the individuals dark-colored because they are members of a population that lives on black soils? Clearly cause and effect can, and often do, operate both ways. Salthe (1985) and others have pointed out that this is a common feature of hierarchical biological organization.

THE CONCEPT OF ECOLOGICAL NICHE

Hutchinson's Concept of Niche

The concept of niche provides a way of characterizing important ecological attributes of species while recognizing their uniqueness. Grinnell's (1917) original concept of niche emphasized the limits on habitat and geographic distribution of species (see James et al. 1984; Schoener 1988). Hutchinson (1957) redefined and formalized the concept of niche. He suggested that the niche of any species could be represented quantitatively in terms of the multidimensional combination of abiotic and biotic variables required for an individual to survive and reproduce, or for a population to persist.

Hutchinson's concept of niche is rooted in two well-established principles of population biology. (1) Malthus's (1798) law of population growth states that all populations have the inherent capacity to increase exponentially in favorable environments. This increase is prevented, and abundance and distribution are ultimately limited, when abiotic conditions, the effects of other species, or the density of other individuals of the same species sufficiently curtail survival and reproduction. (2) Gause's (1934) principle of competitive exclusion states that if two species coexisting in the same environment have identical requirements, one will prove superior in producing offspring, which will eventually lead to the extinction of the other species. The corollary is that species that frequently coexist in the same environments tend to differ in their requirements (MacArthur 1958, 1972b; Hutchinson 1959).

Hutchinson's concept of niche thus implicitly stresses the uniqueness of species in their ecological relationships. The fact that species exhibit distinctive patterns of abundance and distribution reflects their different requirements for environmental conditions. By Hutchinson's operational characterization, the niche is an attribute of species, not of environments. Thus, although extinct species had niches, there can be no unfilled niches.

The concept of the niche also has important evolutionary implications. The niche is dynamic, changing over time and space as the species evolves. Each of the niche variables potentially represents an agent of natural selection. Any heritable change that alleviates the effect of a limiting factor (so long as there is no trade-off that causes a compensatory increase in the limiting effect of other niche variables) will result not only in population growth but also in increased representation of the advantageous genotype in subsequent generations. Thus, the niche concept provides a way of formalizing the relationship between the environmental agents of selection and adaptive microevolutionary change. This concept was one of the promising contributions of the evolutionary approach to ecology developed by Hutchinson, MacArthur, Levins, and others. It was seemingly lost sight of in the debates about the usefulness of certain mathematical models and the importance of interspecific competition that occurred during the last two decades.[4]

Application

There seems to be a feeling among contemporary ecologists that the concept of niche has only heuristic value: it provides a useful way of thinking about the ecological relationships of populations, but is impossible, or at least impractical, to measure. Certainly I know of no case in which an investigator has attempted to measure all of the factors, both abiotic and biotic, that limit the abundance and distribution of any species population.

Although it would be a major undertaking, it should be both practical and rewarding to measure carefully the niches of at least a few species. This endeavor probably would require the collaboration of physiological ecologists, who would work out the limits set by tolerance of abiotic conditions, and population and community ecologists, who would study the effects of intraspecific and interspecific interactions. It would answer a number of important questions: Why is the species restricted to certain habitats and to a particular geographic region? Why does the abundance of the species vary over time and space? How many independent environmental factors impose the primary limits on abundance and distribution, what are these niche dimensions, and on what stages of the life history do they have their effect?

---

4. Schoener (1988) and others have suggested that Hutchinson's concept of niche has been supplanted by the concept of "resource utilization functions" (see also MacArthur 1972b). As they have often been used, such resource utilization functions typically have three components: food, space, time. This concept reflects the emphasis that many evolutionary ecologists placed on biotic interactions, especially interspecific competition. However, I think that a broader concept—one closer to Hutchinson's that emphasizes the role of limiting factors, both abiotic and biotic—is needed to address a range of issues from the structure and dynamics of local populations and communities to the limits of geographic ranges and large-scale patterns of diversity.

Most often ecologists have measured only one or a few related niche variables and ignored others. Physiological ecologists have focused on abiotic conditions that affect the survival and reproduction of individuals by causing physiological stress. They have been particularly successful in identifying some limits to local and geographic distributions of species set by physical conditions (e.g., Porter and Gates 1969; Steenbergh and Lowe 1976, 1977; Nobel 1980; Tracy 1982, 1991; Ehleringer 1984a,b; Root 1988a,c; but see Repasky 1991). Population and community ecologists have most often focused on the effects of intra- and interspecific competition, predation, mutualism, or (more recently) parasitism and disease. They have often showed how one of these biotic interactions influences local abundance and/or distribution (e.g., Connell 1961b; Brown 1971b; Paine 1974; Lubchenco and Menge 1978). We now have excellent examples of a single abiotic or biotic factor that limits a local population of a given species over a short period of time. But we have few examples of the combined effects of many variables on a single species over a long period of time or over the entire geographic range.

It is interesting that Connell's (1961a,b) classic experimental investigation of the barnacle *Chthamalus stellatus* still represents perhaps the most complete study of the niche of any species. Connell's investigations were limited: to the sessile stage of the life history, to the vertical distributional range of the species within the intertidal zone, and to less than three years of field study on one small expanse of rocky shore in Scotland (a tiny fraction of the species' geographic range). Nevertheless they are very illuminating, because they show that abiotic conditions, competition, and predation all play important roles in limiting the local distribution of this species. The niche comprises multiple, relatively independent variables, as Hutchinson's model suggests.

Macroscopic Implications

As limited as the empirical studies of the niche are, they offer valuable insights into the role of ecological processes in biogeographic and macroevolutionary phenomena. First of all, the limits of species' geographic ranges are set by ecological factors. The limits of distribution occur where one or more niche variables so reduce survival and reproduction that they prevent the occurrence of individuals and the maintenance of populations. Furthermore, the boundary of the geographic range shifts over time as environmental conditions change or as the species evolves different requirements.

What biogeographers call barriers are simply unfavorable ecological conditions that are so widespread in space and so constant in time that they are normally sufficient to prevent the dispersal of individuals. There is evidence that a species can become established in a region where it does not presently occur if it is able to disperse across these barriers; an example is the establishment of an exotic species after being transported by humans. When barriers prevent a species from occurring in otherwise habitable areas, they appear to be a legacy of history. They reflect the effects of past geologic, climatic, or oceanographic conditions on the present geographic distribution.

Second, the geographic patterns of diversity reflect the influence of the earth's environment, both present and past. High species richness within small local areas (what ecologists call high alpha diversity) must reflect both present conditions that satisfy the different requirements of the many coexisting species, and past conditions that have enabled these species to accumulate as a result of some combination of colonization from other regions and speciation within the region. Similarly, rapid turnover of species from place to place across the landscape (what ecologists call high beta diversity) must reflect underlying spatial variation in the environment.

It is important to emphasize at this point that species range boundaries and geographic patterns of diversity ultimately reflect variation in the abiotic environment. Biotic interactions such as competition, predation, and mutualism can be proximate limiting factors, but it is the physical template of the earth—its geology, climate, and physical oceanography and limnology—that ultimately determines the distribution of life.

To see this, consider the example of Connell's barnacles (fig. 3.1). Connell clearly showed that competition from another barnacle species, *Balanus balanoides,* largely determines the lower distributional limit of *Chthamalus stellatus* in the intertidal zone. When Connell experimentally removed *Balanus, Chthamalus* was able to become established and persist substantially below its normal range. But why does the upper limit of *Balanus* occur at a particular tidal height? Connell's experiments provided the answer: the physiological stress of exposure between tides determines the upper distributional limit of *Balanus,* and thus abiotic conditions ultimately determine where competition proximally limits the range of *Chthamalus.*

Sometimes the effects of the earth's abiotic template reflect past, rather than present, conditions, and thus appear to be a legacy of history. To see this, imagine the following situation. Two lakes, one in British Columbia and the other in Ontario, have virtually identical limnological conditions (temperature, depth, water chemistry, etc.), and very similar kinds and numbers of phytoplankton and zooplankton, but the lake in Ontario has only one-fourth the number of fish species that inhabits the lake in British Columbia. One promising hypothesis to account for the difference

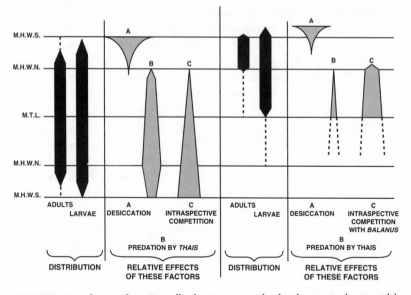

FIGURE 3.1. A diagram from Connell's classic paper on the distributions in the intertidal zone of adults and newly settled larvae of two barnacle species, *Balanus balanoides* and *Chthamalus stellatus,* showing the relative importance of three features of the niche: desiccation during exposure between tides, predation by the gastropod mollusk *Thais,* and competition from their own and the other species. The letters on the left indicate the position with respect to the tidal cycle, with M.T.L. being mean tidal level, M.H.W.S. being mean high water of spring tides, and so on. In my mind, Connell's study warrants recognition not only as a pioneering application of manipulative field experiments in ecology, but also because it remains one of the most thorough studies of the niche of any species. (From J. H. Connell, "The Influence of Interspecific Competition and Other Factors on the Distribution of the Barnacle *Chthamalus stellatus,*" *Ecology* [1961] 42:722, fig. 5. Copyright (c) 1961 by the Ecological Society of America. Reprinted by permission.)

in fish diversity would be that the lake in Ontario is in an area where there was glaciation during the Pleistocene. The glaciers prevented the occurrence of all aquatic organisms until 10,000 years ago. Since then the phytoplankters and zooplankters, which have resistant stages capable of long-distance overland dispersal, have recolonized, but only a few species of fish, which require aquatic connections in order to disperse, have been able to reestablish populations. By contrast, the lake in British Columbia is in an unglaciated region and has been capable of supporting fish populations for hundreds of thousands of years; over this time period some combination of colonization from outside and speciation within the lake has produced a much more diverse fish fauna than in the lake in Ontario. While this example is hypothetical—it would be very difficult to find nearly identical lakes in regions with such different histories—it has much in common with many real situations (e.g., Tonn et al. 1990) in which differences in the composition and diversity of at least some components of the biota reflect the influence of past rather than present abiotic conditions.

A third macroscopic implication of the niche concept concerns the role of the environment in the macroevolutionary processes of species dynamics. The fossil record documents dramatic changes in the diversity of life on earth as a consequence of episodes of speciation and extinction (e.g., Stanley 1979, 1982, 1986). For example, Raup (1979, 1986) estimates that as many as 96% of the existing marine species may have gone extinct at the end of the Permian, 250 million years ago. What environmental event caused this catastrophe, and why were some lineages able to survive when so many perished? Phylogenetic reconstructions of taxonomic groups typically reveal that some lineages have speciated to leave many surviving descendants, whereas other clades either failed to diversify or their descendant species have nearly all gone extinct (Donoghue 1989, Brooks and McLennan 1991). Were the successful lineages those that "invented" special combinations of traits, and if so, how did these traits alter interactions with the environment so as to promote population growth, geographic expansion, and speciation (for one example, see Vrba 1992)? Addressing these issues will require information on the nature of past environments and the evolution of the niches of the species that inhabited them.

THE ORGANIZATION OF COMMUNITIES

Gleason's Concept of Community

In the early part of the twentieth century, two eminent plant ecologists, F. E. Clements (1916, 1949) and H. A. Gleason (1917, 1926), debated the nature of what they called "plant associations." As in most such arguments in ecology, both protagonists were largely right; they were just talking about different things. Clements emphasized the emergent properties of communities, especially the predictable patterns and processes of secondary succession that occurred over long periods and large regions. Gleason focused on the idiosyncratic details, especially those revealed by differences in the presence, absence, and relative abundances of species, even when nearby sites with similar environments were compared. We can now recognize that Clements and Gleason were focusing on two different features of complex systems: whole systems exhibit common emergent properties despite the enormous variation in the detailed structure and dynamics of their components.

Much of the remainder of this book is concerned with the emergent properties of species associations. Here I make the point that these regu-larities are superimposed on what I call a "Gleasonian individualism." In Gleason's view, the community was an opportunistic collection of species: those that had the capacity to disperse to a site and become established there. This view of the species composition of a community is a logical consequence of the Hutchinsonian niche concept. Given that (1) each species has unique requirements that determine its distribution and abundance, and also has a history of relationships with biogeographic bar-riers that may further limit its distribution to only a subset of the places where it could potentially occur, and that (2) environmental variation is so great that no two places have exactly the same present conditions or past histories, then no two places should be inhabited by exactly the same assemblage of species. Even when the species inhabiting two sites are very similar in taxonomic composition, there invariably are distinctive dif-ferences in the relative abundances, spatial distributions, and patterns of temporal fluctuation of the component species.

This view of ecological communities, derived from Hutchinson's con-cept of niche and Gleason's view of plant associations, is well supported by the data. Whittaker (1956, 1960; Whittaker and Niering 1965), in a number of intensive studies of the distribution and abundance of plants along environmental gradients in several areas of North America, repeat-edly showed that species composition varied widely and essentially con-tinuously, reflecting the unique niche requirements of the individual species.

Similar conclusions follow from my own studies of spatial and temporal variation in the composition of small mammal assemblages in the deserts of southwestern North America. When many different sites of generally similar habitat were censused, nearly every site was inhabited by a unique combination of species, and when exactly the same set of species did oc-cur at more than one site, their relative abundances were different (fig. 3.2; Brown and Kurzius 1987). Furthermore, when the community inhab-iting my 20 ha experimental site in the Chihuahuan Desert was monitored carefully over many years, the relative and absolute abundances of the species changed continually; there were even changes in species composi-tion owing to local colonization and extinction events (fig. 3.3; Brown and Kurzius 1989; Brown and Heske 1990b).

Community Structure

Much of the last three decades of community ecology has been devoted to the analysis and interpretation of "community structure." To a large extent this emphasis reflects the influence of Hutchinson and his student

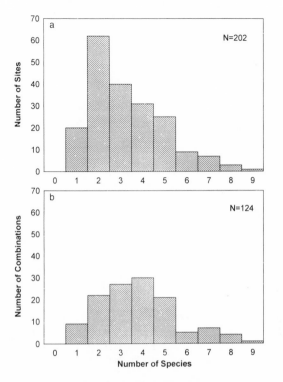

FIGURE 3.2. Frequency distributions of (a) the number of sample sites and (b) the number of different combinations of species inhabiting those sites as a function of species richness in seed-eating rodent communities in relatively homogeneous desert shrub habitat in the southwestern United States. Even though there were only twenty-eight species in the pool, and two to five of these occurred at most sites, the vast majority of the sites (124 out of the 202) supported a unique combination of species. The most frequently observed combination, of two species, occurred only six times. (Adapted from Brown and Kurzius 1987.)

MacArthur, who noted that the closely related species that coexisted in local ecosystems frequently differed conspicuously in attributes such as body size, size of trophic appendages (e.g., bird bills), and use of microhabitats (e.g., MacArthur 1958, 1972b; Hutchinson 1959). These differences were hypothesized to reflect differences in resource use that reduced interspecific competition and promoted coexistence. Actually, several earlier ecologists had noted this kind of community structure (e.g., Grinnell and Orr 1934; Lack 1947). Brown and Wilson (1956) had suggested that competition among coexisting species tended to result in the evolution of greater differences between sympatric populations than between allopatric ones, a phenomenon that they termed character displacement.

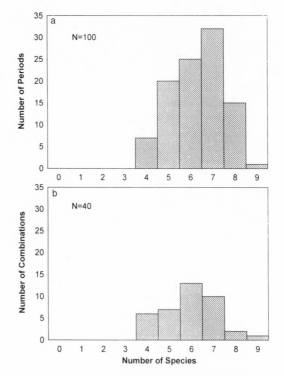

FIGURE 3.3. Frequency distribution of (a) the number of sample periods and (b) the number of different combinations of species observed during those periods as a function of species richness of seed-eating rodents occurring at my 20 ha study site in the Chihuahuan Desert of Arizona during periods of monthly trapping from 1977 to 1986. Although the pool contained only ten species, there was enormous variation in which ones were actually present from month to month and year to year; forty different combinations were observed over the 10 years. While some of the absences can be attributed to failure to capture individuals that were present on the site (especially in the case of two pocket mouse species that hibernated), the vast majority of the variation in species composition was owing to local metapopulation colonization/extinction dynamics. (Adapted from Brown and Kurzius 1989; see also Brown and Heske 1990b.)

The last ten or fifteen years have been marked by much debate, much of it unfortunately rancorous, about the nature and causes of such "structure" in ecological communities (e.g., Salt 1983; Strong et al. 1984). This debate has had two beneficial effects, however. First, it has resulted in the development of rigorous statistical criteria for the recognition of structure. Community structure has come to be defined as nonrandom characteristics of locally coexisting species. Structure as reflected in some trait, such as body size or bill length or foraging height, can be demonstrated by showing that it is either (1) distributed differently within a single set of coexisting species than expected by chance (fig. 3.4), or (2)

SONORAN DESERT   GREAT BASIN DESERT

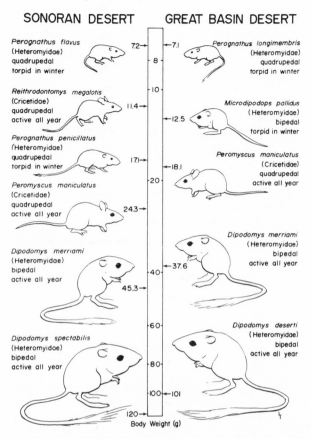

FIGURE 3.4. Patterns of distribution of body sizes in two communities of seed-eating desert rodents, one from the Sonoran (left) and the other from the Great Basin Desert (right), which might appear to reflect character displacement. While the species in both assemblages differ substantially in body weight, when analyzed separately by Simberloff and Boecklen (1981) using the Barton and David test statistic, only the Sonoran community was significantly more regularly spaced on the logarithmic scale (reflecting a more constant ratio of sizes among adjacent species) than expected by chance. (From Brown 1975; figure reprinted by permission of the publishers from *Ecology and Evolution of Communities*, edited by Martin L. Cody and Jared M. Diamond, Cambridge, Mass.: The Belknap Press of Harvard University Press, Copyright (c) 1975 by the President and Fellows of Harvard College.)

distributed differently in multiple sets of coexisting species than expected if the species were assembled at random from some appropriate regional species pool (fig. 3.5).

The other beneficial effect of the debate is that it has stimulated the use of well-designed, replicated field experiments to assess the roles of biotic interactions in regulating the local abundance and distribution of species, and thereby in influencing community structure. Many of these

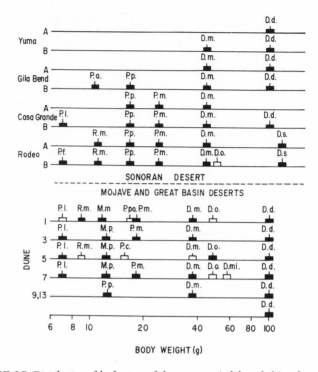

FIGURE 3.5. Distribution of body sizes of the common (solid symbols) and rare (open symbols) species of seed-eating desert rodents that coexisted at fourteen geographically separated sites in the Great Basin and Sonoran Deserts. When these communities were tested collectively, they exhibited much more regular spacing on the logarithmic axis (more constant size ratios among adjacent species) than expected by chance (Bowers and Brown 1982; Hopf and Brown 1986). Because of differences in the power of the statistical tests that can be applied, traits that cannot be shown to exhibit statistically regular spacing within any one community (such as the Great Basin rodent community in fig. 3.4) may exhibit highly significant character displacement when comparisons are made across many communities. (From Brown 1975; figure reprinted by permission of the publishers from *Ecology and Evolution of Communities,* edited by Martin L. Cody and Jared M. Diamond, Cambridge, Mass.: The Belknap Press of Harvard University Press, Copyright (c) 1975 by the President and Fellows of Harvard College.)

experiments have unequivocally demonstrated that not only interspecific competition, but also predation and mutualism, affect the composition and dynamics of communities, and that these interactions vary in space and time (Menge et al., in press).

Now the debate about community structure has largely died away. Even some of the most vocal skeptics, such as Daniel Simberloff (e.g., compare Connor and Simberloff 1979; Simberloff and Boecklen 1981 with Dayan, Simberloff et al. 1989, 1990, 1992; Dayan and Simberloff 1994), now seem to agree that many communities are composed of species that tend to be more different in their morphology and/or behavior than expected by chance, and that interspecific competition is often, but by no means always, the most likely explanation for this pattern.

It is now clear that assemblages of coexisting species can, and often do, exhibit several different kinds of structure. One, already mentioned, is the tendency toward uniform (or in some cases aggregated, suggesting some kind of character convergence rather than character displacement) dispersion of species traits. Another kind of structure is the tendency of communities to exhibit nonrandom taxonomic or ecological attributes as reflected in "assembly rules" (Diamond 1975). For example, Barry Fox (1989; see also Fox and Kirkland 1992; Fox and Brown 1993) has shown that Australian small mammal communities can be characterized by the rule that each different genus (which also represents a different ecological guild) tends to be represented by one species before a second species of any genus is added; two species of each genus are present before a third species is added; and so on (fig. 3.6). A third kind of structure is the tendency of species to be distributed among assemblages as "nested subsets": that is, communities of successively lower species richness tend to be subsets of more diverse assemblages (table 3.2; May 1976; Brown and Gibson 1983; Patterson and Atmar 1986; Patterson 1987, 1990; Kodric-Brown

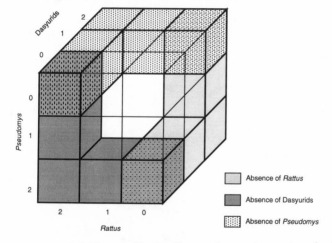

FIGURE 3.6. Fox's rule for the assembly of small mammal communities in Australian heathlands, based on the premise that species in different functional and taxonomic groups are more likely to be able to coexist. The rule states that species are added to communities such that one in each genus is present before a second species in any genus is added, and so on. Thus communities can be characterized in terms of species combinations favored (unshaded) and unfavored (shaded) by the rule, and all possible combinations of zero to two species per genus can be represented by the 3 x 3 cube. Real communities are predicted to be represented by more favored and fewer unfavored combinations than expected by chance. (Adapted from Fox 1989.)

TABLE 3.2 Nested subset structure of the communities of fishes inhabiting twenty-eight isolated springs in the Dalhousie Basin of Central Australia

Species	A 1 2	CA 5	CC 3	CD 1	CD 2	E 1	CB 1 2	CC 3 2	A 3	B 1 2	CD 3	GA 1 2 3	GA 6	DA 3	DB 2	DB 3	E 5	GB 2	A 5	CA 11	CC 4	DA 2	DB 1	E 2	F 2	GA 5	H 3	No. of species
Goby	X	X	X	X	X	X	X	X	X	X	X	X	X	X	X	X	X	X	X	X	X	X	X	X	X	X	X	28
Gudgeon	X	X	X	X	X	X	X	X	X	X	X	X	X	X	X	X	X	X										19
Catfish	X	X	X	X	X		X	X	X	X	X	X	X															14
Hardyhead	X	X	X	X	X		X	X																				9
Perch	X	X	X	X		X																						7
Number	5	5	5	5	4	4	4	4	3	3	3	3	3	2	2	2	2	2	1	1	1	1	1	1	1	1	1	

Source: Kodric-Brown and Brown 1993a.

Note: In this nearly perfect nested subset structure, as the number of species increases from one to five, species are added in a predictable order so that each spring with higher species richness tends to have all of the species present in less diverse communities plus one additional species. There is only one exception to this pattern, the absence of hardyhead from Spring E1. The nested subset structure implies that highly predictable colonization/extinction processes ultimately determine the composition of these communities.

and Brown 1993a). A related, but different, kind of structure is what Hanski (1982a,b,c) calls core-satellite organization: the tendency of assemblages to be composed of two classes of species, some that are abundant and found in many sites and others that are rare and restricted to a few sites. At least the first two of these kinds of structure suggest the influence of interspecific competition, which should be especially strong among, but not necessarily confined to, species that are closely related or otherwise similar in morphology, physiology, and behavior.

Different kinds of community structure are likely to be caused by the other kinds of biotic interactions: predation (including herbivory, parasitism, and disease) and mutualism. Compared with the effects of competition, the kinds of structure expected to result from these interactions are difficult to characterize. In order to avoid competition, species must differ in some way in their use of limiting resources, but it is less clear what kinds of traits will enable species to capture their prey, escape their enemies, or obtain benefits from their mutualists. This is probably the reason why so much of the attention to community structure has been focused on competition. There are, however, good examples of community structure caused by predation and mutualism. Gareth Vermeij (1978, 1987) presents a convincing case that predation on gastropod mollusks has led to the thickening and "ornamentation" of their shells. The literature of plant-herbivore and plant-pollinator interactions contains many examples of morphological and physiological traits of coexisting species that presumably can be attributed to the interactions between them (e.g., Ehrlich and Raven 1965; Futuyma and Slatkin 1983).

Reconciling Gleasonian Individualism with MacArthurian Structure

A major problem now facing community ecologists is how to reconcile the individualism of species documented by Gleason, Whittaker, and others with the kind of community structure emphasized by MacArthur, Hutchinson, and others. They are not incompatible. Gleason (1917, 1926) recognized the importance of interactions among species in affecting the distribution of species and the composition of assemblages. MacArthur (1972b) presented examples to show that the distributions and abundances of individual species were limited by seemingly idiosyncratic interactions with the abiotic environment or with other species. This is another case of complex systems exhibiting enormous variation in the details of their components at the same time that they also exhibit common, emergent features.

Nevertheless, there remains a tension between the Gleasonian and MacArthurian perspectives. It is sufficient to make two points. The first is that it is incorrect to equate Gleasonian individualism with the influence of abiotic conditions and MacArthurian structure with the effects of biotic interactions. The interactions of a species with other organisms are just as individualistic as its relationships with the physical environment because they depend ultimately on the same unique features of morphology, physiology, and behavior. Furthermore, the abiotic environment can cause patterns of community structure just as pronounced as those caused by interspecific interactions. Excellent examples are provided by several studies of structural and functional convergence among distantly related organisms inhabiting similar environments on different continents (e.g., Fuentes 1976; Mooney 1977; Cody and Mooney 1978). Interestingly, most such studies draw both MacArthurian and Gleasonian conclusions.

For example, Mooney et al. (1977; Cody and Mooney 1978) document dramatic similarities in growth form, leaf morphology, and physiological attributes among plants inhabiting Mediterranean climatic regions in southern Europe and northern Africa, California, Chile, South Africa, and southwestern Australia. These similarities reflect similar adaptations to the cool, rainy and foggy winters, the hot, dry summers, the shallow, often nutrient-poor soils, and the frequent, intense wildfires in all of these regions. Almost as striking as the convergences in vegetative structural and functional attributes, however, are the differences among these plants in their flowers and seeds. These differences reflect divergent adaptations to exploit different kinds of animals for pollination and seed dispersal. Note that in this case the MacArthurian structure is caused largely by abiotic conditions, while the Gleasonian individualism can be attributed to adaptations to biotic interactions. Furthermore, when convergence is apparent, it is in the characteristics of the communities as a whole, rather than in the individual species.

The second point has to do with how fine-tuned are the ecological and especially the evolutionary processes that structure communities. Even if species are individualistic and opportunistic, ecological interactions can play a major role in determining which species coexist at a local site by influencing the probabilities of immigration and establishment. Janzen (1985; see also Brown 1989; Hopf, Valone, and Brown 1993) called this ecological sorting. It is also true, however, that a substantial period of close association is probably required for coevolutionary adjustments among coexisting species.

There are examples to support the entire spectrum between these two extremes. There is abundant evidence, especially from human-caused perturbations, for opportunistic invasions of exotic species and rapid changes in community composition. For example, Strong (1974; Strong, McCoy, and Rey 1977) and Connor et al. (1980; see also Southwood 1978; Southwood, Moran, and Kennedy 1982) document the rapid buildup of communities of mostly native insect herbivores on introduced plant species. On a slightly longer time scale, Graham (1986) describes major changes in small mammal assemblages since the late Pleistocene. Species that occurred together only 20,000 years ago currently have geographic ranges separated by hundreds of kilometers, and species that presently occur together were widely allopatric during glacial periods. It is hard to imagine how fine-tuned, species-specific coevolutionary relationships could develop and persist in the face of such spatial and temporal variation. If such examples are the rule, then during the last few decades many evolutionary ecologists—myself included—probably overestimated the degree to which communities were structured by coevolutionary adaptations (see also Futuyma and Slatkin 1983).

On the other hand, there are also examples of close coevolution between pairs of coexisting species. There are orchids whose flowers mimic the morphology and pheromones of females of certain insect species in order to trick the males into visiting and pollinating them. Both birds (crossbills, genus *Loxia*) and mammals (red squirrels, genus *Tamiasciurus*) that feed on conifer seeds have special morphological adaptations for opening the cones of particular tree species (Smith 1970; Benkman 1993).

It is probably no coincidence that some of the best-documented cases of finely tuned, species-specific coevolution come from small oceanic islands. Examples include character displacement in Galápagos finches and Antillean *Anolis* lizards (e.g., Lack 1947; Schoener 1970; Grant 1986; Roughgarden, Heckel, and Fuentes 1983). One feature of such islands is that they have few species, which limits the possibilities for interspecific interactions. For a species of Darwin's finch inhabiting a small island, the interspecific competitors are highly predictable. Throughout the island and throughout the year, the most severe competitor is likely to be another finch species. Thus it is easy to imagine how a change in bill size might increase fitness by reducing interspecific competition for food. For a species of sparrow inhabiting western North America, however, the competitors are much less predictable. For one thing, there are several other species of sparrow with geographic ranges overlapping its range, but the abundance, distribution, and identity of these potential competitors vary spatially, among habitats and from region to region, and temporally, among seasons and from year to year. There are also many other kinds of organisms (e.g., mammals and insects) that occur together with the sparrow and potentially compete with it for food. It is much harder to imagine how selection might act on bill size in this case.

Ever since Darwin and Wallace, islands have provided model systems for ecological and evolutionary studies. But to what extent can the patterns and processes documented in small, relatively closed systems, such as islands, lakes, and caves, be assumed to hold in large, open systems, such as continents and oceans—and vice versa? I suspect that for many ecological and evolutionary phenomena, tiny, isolated oceanic islands or lakes represent opposite ends of a spectrum from large continents or oceans. Neither of these extremes provides a very good model for the other.

CONCLUDING REMARKS

This chapter presents some background that will be used throughout the remainder of the book. The definitions and concepts of species, niche, and community remain the subject of considerable discussion and debate. While I recognize the existence—and some of the merits—of the different points of view, I have tried to steer clear of the controversies. Instead, I have tried to use operational definitions, to suggest how the differences might be reconciled, and to develop a simple, realistic conceptual framework that can be used to make connections between traditional ecological and microevolutionary studies and the macroscopic phenomena that will be considered in subsequent chapters.

Central to the entire book is the recognition that biological diversity is organized. Variation among individual organisms in morphology, physiology, behavior, genetics, and ecology is neither continuously nor randomly distributed; it is clumped into units with gaps separating the clumps. These units are called species. This discreteness of attributes not only permits species to be identified, but also reflects the history and process of species formation and the unique ecological roles that enable species to persist and to coexist with other species in ecological communities.

4

The Abundance and Distribution of Species

Imagine that all the individuals of one species could be made to glow so brightly as to be visible from space. If we could take a photograph from a satellite of the entire geographic range of the species, all of the individuals would appear as little points of light. Such a photograph would capture many interesting features of abundance and distribution. Further extensions of this thought experiment that would provide additional data immediately suggest themselves. By taking multiple time-lapse photographs, we could measure movements and dispersal. By coding age and sex classes or genetic variants with different-colored lights, we could monitor the spatial and temporal dynamics of population structure.

This not so futuristic study (see Rapoport 1982, his figure 6.3) would produce a wealth of information. It would provide graphic images to reinforce things we already know: that most higher plants move only as seeds, and that the geographic ranges of most species contain within their boundaries large "holes" of uninhabited area. More importantly, it would provide information that presently is unavailable: on the paths taken by successful and unsuccessful dispersers, and on the structure and dynamics of local populations in different parts of the range.

THE MEASUREMENT OF ABUNDANCE AND DISTRIBUTION

One thing that our photographic technique would make explicit is the relationship between abundance and distribution. The photographs would show that abundance or population density is just a crude statistical representation of the spatial distribution of individuals. The fundamental units are the individuals, and at any instant in time these can be represented as points in space. Measuring population density requires choosing some arbitrary boundaries, counting the individuals within them, and

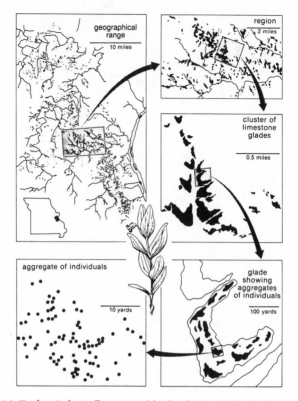

FIGURE 4.1. Erickson's classic illustration of the distribution (in black) of the plant *Clematis fremontii* on different spatial scales within the state of Missouri. These maps are largely characterizations of the actual patterns of distribution, because it would be a Herculean task to obtain the enormous quantities of direct data needed to plot distributions down to the scale of individuals. (From Erickson 1945, fig. no. 4 on p. 423.)

then dividing by the area. This method gives one number, N individuals per hectare, at the expense of discarding much of the information embodied in the distributional data.

Furthermore, as shown in figure 4.1, population density determined in this manner is a very crude statistic, and one that can actually be misleading. It does not take account of the variation in spacing among individuals, and consequently it is insensitive to spatial patterns of dispersion.[1] The photographic thought experiment makes clear that any other measurement of abundance that we might make—for example, by a visual

1. Of relevance here is a growing literature on how to deal conceptually with problems of scale. For a general treatment, I refer readers to a recent paper by Levin (1992). A specific example is the intriguing suggestion by Maurer (1994, in press) that the geographic range of a species might represent a fractal distribution of individuals. For a brief introduction to fractals, see chapter 5.

census or a mark-recapture study—also provides just one, very imperfect, measure of distribution. Nevertheless, even very simple measures of spatial distribution provide much more information than does population density alone. For example, if we were to measure the nearest-neighbor distances between all the individuals, we could easily calculate density from the mean nearest-neighbor distance, but we would also obtain additional, valuable data on the variance in the spacing of individuals.

The photographic technique would make it equally clear that a map of the geographic range of a species is also a very imperfect representation of the distribution of individuals. Available maps of geographic ranges are usually of two types: the "dot map," in which the points represent all known locations where individuals of the species have been known to occur (usually those sites where museum specimens have been collected); and the "boundary map," in which a free-form line has been drawn to enclose the peripheral records of occurrence, creating some irregular area. Although the dot map conveys slightly more information than the boundary map, neither begins to capture the richness of information contained in the photograph. Even with the dot map, we do not know whether the blank spaces represent areas where the species does not occur or just areas that have not been sampled adequately; also, while the dots do indicate where the species was present, they do not convey any information about the spatial or temporal variability in its abundance at those localities.

Actually, the geographic range of a species is a very complex and dynamic reflection of the distribution of all the individuals. Rapoport (1982) has pointed out that the range can be thought of as a slice of Swiss cheese, containing holes of varying sizes and shapes within its irregular outer boundary. An increasing number of maps use census data and isopleths to show the distribution of some estimate of abundance over the range. But even these are gross simplifications. The real units of distribution are individuals, and like the glowing lights in the hypothetical satellite photographs, they are dynamic in space and time. They vary in their distances from each other, move around, disperse into previously vacant areas, disappear from places where they were formerly present, and wink on and off as births and deaths occur.

The reason for spending so much time on this thought experiment and methodological considerations is to raise a flag of caution before proceeding. Unfortunately, time series of maps giving the locations of individuals within large geographic regions are simply not available for most organisms over any extensive area (but note that Rapoport [1982] was able to plot the locations of mature individuals of a palm species because of their distinctive appearance in aerial photographs). For the foreseeable future, studies of abundance and distribution will have to use the limited data available: the estimates of population density made by ecologists and the geographic range maps prepared by biogeographers and systematists. Although these data are only a crude representation of the wonderfully complex and dynamic distributions of individual organisms, much can be learned from them.

THE RELATIONSHIP BETWEEN ABUNDANCE AND DISTRIBUTION WITHIN SPECIES

Pattern

So what would one of our satellite photographs showing the instantaneous distribution of the individuals of a species over geographic space look like? We know that the dots would not be dispersed uniformly; they would be densely clustered in some places, sparsely dispersed in others, and absent altogether from others. But would there be any general pattern that tends to be repeated across different species?

Several empirical studies suggest that most of the distributions would resemble the deposition of soot particles from a smokestack on a calm day. The average concentration of individuals tends to decrease from the center of the range toward the margins, while the size of uninhabited patches shows the opposite trend. This pattern was perhaps first quantified by Robert Whittaker (1956, 1960, 1967; see also Whittaker and Niering 1965) in his classic studies of the distribution of plants in ecological gradients on mountainsides. Whittaker fitted his data with normal curves, which suggested that abundance was highest at the center of each species' range and declined gradually and symmetrically toward the boundaries. Hengeveld and Haeck (1982) repeatedly observed a decline in abundance from the center to the margin in transects across the geographic ranges of several species of European plants and insects (fig. 4.2). Contour maps of density of both breeding and wintering North American bird species show similar patterns (fig. 4.3; Bystrak 1979; Brown 1984; Robbins, Bystrak, and Geissler 1986; Root 1988a,c). Rapoport (1982), in his analysis of the Argentine palm (*Coperinicia alba*) referred to above, noted that as the edge of the geographic range was approached, not only did abundance decline but also the distribution became more patchy, leaving increasingly large areas uninhabited.

Hypothesis

In 1984, I suggested a relatively simple explanation for this pattern of variation in abundance over the geographic range. It was based on three assumptions:

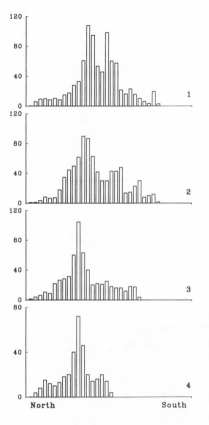

FIGURE 4.2. Abundance of the plant *Avena sativa* in four north-south transects across its geographic range in Russia. Note that the species tends to reach its highest local population densities at sites near the center of its distribution within the latitudinal gradient. (Adapted from Hengeveld and Haeck 1982, fig. no. 6 on p. 311.)

1. Abundance and distribution reflect the response of local populations to local conditions.

2. Local abundance and distribution reflect the extent to which local environments meet the multiple Hutchinsonian niche requirements of each species.

3. Environmental variables that affect abundance and distribution tend to be autocorrelated over space, so that sites in close proximity tend to have more similar abiotic and biotic characteristics (including interspecific interactions) than more distant sites.

I will return to these assumptions shortly, but let's accept them for the moment. From the first two it follows that abundance should vary over the range, from high values where local abiotic and biotic conditions are most conducive to survival and reproduction, to zero where one or more variables preclude occurrence. The third assumption implies that abundance will be spatially autocorrelated because the determining environmental variables are. In the very simplest case, where all limiting environmental variables change continuously and monotonically across the

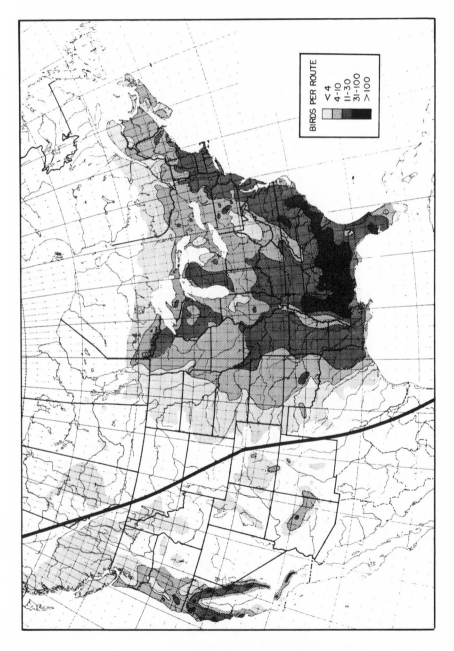

FIGURE 4.3. A map showing spatial variation in population density over the geographic range for two closely related bird species, the relatively continuously distributed blue jay (*Cyanositta cristata*, to the right of the heavy line) and its more patchily distributed congener, Steller's jay (*C. stelleri*, left of the heavy line). These data, from the North American Breeding Bird Survey, show the extent to which peaks of abundance tend to be located centrally within the range. (From Robbins, Bystrak, and Geissler 1986.)

BIRDS PER ROUTE

< 4
4-10
11-30
31-100
>100

landscape, these assumptions predict that abundance will be highest in one most favorable site near the center of the range and will decline gradually toward all the boundaries (fig. 4.4a). This is exactly the pattern that Whittaker implied when he used normal curves to show variation in abundance in transects across the elevational ranges of plant species.

It is possible to imagine exceptions to this pattern. Obvious exceptions will occur if the variation in limiting environmental factors is either discontinuous or multimodal. Discontinuous changes in one or more limiting niche variables can cause rapid shifts in abundance (fig. 4.4b). For example, abrupt changes in either abiotic conditions (e.g., at a land-water interface) or other species (e.g., the presence or absence of a mutualist, predator, or competitor) may change the environment from favorable to totally unsuitable, causing a sharp edge of the range. Multimodal variation in one or more environmental factors can cause multiple peaks and valleys

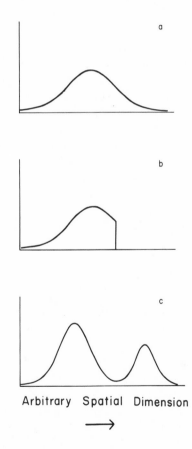

FIGURE 4.4. Hypothetical patterns of abundance within the geographic range. (a) A unimodal, normal-shaped distribution, expected when niche parameters vary as continuous gradients, with spatial autocorrelation on all scales. (b) A truncated distribution, expected if there is abrupt, discontinuous variation in a single, important niche parameter. (c) A multimodal distribution, expected when important niche parameters exhibit a periodic or irregular pattern of spatial variation. (From Brown 1984.)

in abundance across the geographic range (fig. 4.4c). The spatial pattern of abundance may appear either regular, if the environmental variation is periodic (as in the example of basin-and-range topography in the western United States; see fig. 1.1), or irregular, if the environmental variation is more complicated.

The above is a summary of the simple verbal and graphical model that I presented in 1984. Like any model, its ability to account for or predict empirical observations depends on the accuracy of its assumptions and the rigor of its internal logic. The logic appears to be sound. Michael Sanderson and I (unpub.) developed a computer simulation model that incorporated the above assumptions, and it produced the predicted pattern of density declining gradually from the center of the range to the margins (but see below).

The realism of the assumptions is a much more delicate point. All three assumptions are probably valid in a general, qualitative, statistical sense, but it is possible to imagine many specific exceptions. The first assumption, that spatial variation in abundance and distribution reflects responses to local conditions, may not hold when the range of a species is rapidly expanding or contracting (e.g., in the case of colonizing exotic species; see Hengeveld 1989; van den Bosch, Hengeveld, and Metz 1992) or when rates of dispersal are either very high, so that individuals flood into unfavorable places, or very low, so that individuals cannot colonize favorable places. The second assumption, that variation in abundance reflects the extent to which local environmental conditions meet the species' niche requirements, ignores the time lags inherent in population dynamics. The third assumption, of spatial autocorrelation in environmental variables, ignores the extreme degree of spatial heterogeneity, including discrete patchiness, that we often observe in nature.

Evaluation: New Results

It is now time to ask whether this niche-based model is sufficient to account for what we know about abundance and distribution, especially what we have learned from new data and analyses since my 1984 paper. In making such an evaluation, it is important to ask whether alternative hypotheses are equally or more consistent with the empirical observations.

From my own work on intraspecific spatial variation in abundance, mostly done in collaboration with Dave Mehlman and George Stevens (J. H. Brown, D. W. Mehlman, and G. C. Stevens, unpub.), we have learned five important things. First, there are orders of magnitude variation in local population density. We have plotted the data in two ways: as frequency distributions on doubly logarithmic axes and as "Preston

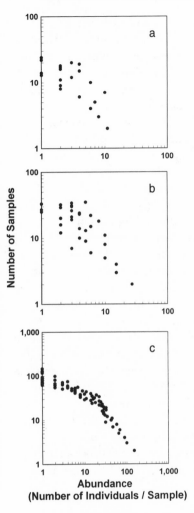

FIGURE 4.5. Frequency distributions of local population abundance among sample sites, showing the extremely heterogeneous spatial distribution of abundance of *(a)* a ciliate parasite, *Trichodina* sp., among host individuals of the fish *Fundulus zebrinus* in Nebraska (data from J. Janovy); *(b)* a breeding bird, the red-eyed vireo (*Vireo olivaceus*), among Breeding Bird Survey sites across its geographic range in North America; *(c)* an annual plant, *Eriogonum abertianum,* among 0.25 m² sample plots in Brown's 20 ha experimental study site in the Chihuahuan Desert. Note that the different kinds of organisms sampled at very different spatial scales exhibit strikingly similar patterns. (From J. H. Brown, D. W. Mehlman, and G. C. Stevens, unpub.)

curves" of frequency on $\log_2$ octaves of abundance (fig. 4.5). Note that these plots show very similar patterns, regardless of kind of organism or spatial scale of sampling;[2] in figure 4.5 I present data for the abundance of a bird over its entire geographic range, of a parasite among host individuals within a geographic region, and of a plant among small sample plots within a few hectares of relatively uniform habitat. Each species appears to be characterized by a few "hot spots" where it is extremely common, many "cool spots" where it occurs but is rare, and many "cold spots"

2. This generalization seems to hold so long as the samples sites are separated by sufficient distance to incorporate local environmental heterogeneity and to insure they include the home ranges of different individuals.

where it is absent. It is surprising that such a seemingly general pattern of variation has not been reported by previous investigators. An exception is the well-documented, highly aggregated distribution of parasites among hosts, which parasitologists have usually fitted with a negative binomial distribution. We have not used a negative binomial, because to do so would require that we include zero values for samples where the species is absent. Including the zeros and fitting a negative binomial would imply that all sample sites are equally suitable for occupation. While this might be justified for parasites—although all hosts might also not be equally susceptible to infestation—it definitely cannot be assumed for the spatially dispersed sites used to sample the other organisms.

Second, even when these zero values are ignored, there is enormous spatial heterogeneity in abundance over the geographic range of a species. Further, much of this variation occurs at very small scales, so that it is not uncommon to find adjacent sample sites in which a species is very rare in one and orders of magnitude more common in the other. In figure 4.6 I show an example, using data from the North American Breeding Bird Survey (BBS). I suspect that this is a general pattern, however, because I found comparable variation when I plotted Whittaker's original data for the numbers of plants in samples along elevational and moisture gradients (fig. 4.7). Whittaker and nearly all other investigators (e.g., fig. 4.3) have used statistical averaging techniques to smooth and interpolate their raw sample data before plotting; these techniques obscure the extremely spiky pattern of spatial variation in abundance.

Third, despite this spatial heterogeneity, there is in the BBS data a pervasive tendency for abundance to be autocorrelated over space, and to be higher near the center than near the edge of the geographic range (figs. 4.6 and 4.8). Furthermore, abundance tends to be higher adjacent

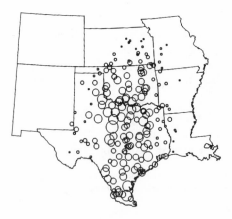

FIGURE 4.6. Spatial pattern of abundance of a breeding bird, the scissor-tailed flycatcher, over its geographic range. Data are from the Breeding Bird Survey. Diameter of each circle is proportional to the average number of individuals counted at each census site. Note the great variation in abundance, with the highest values occurring near the center of the range. (Courtesy of D. Mehlman.)

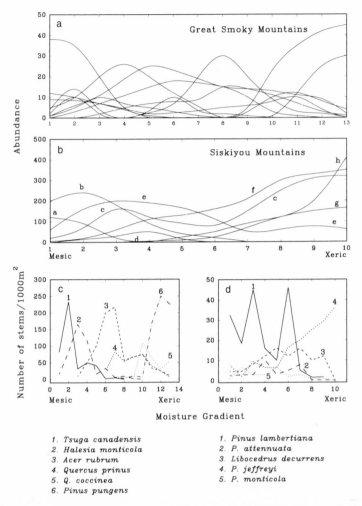

FIGURE 4.7. Distribution of tree species along moisture gradients in the Great Smoky (a, c) and Siskiyou Mountains (b, d) based on data from Whittaker (1956, 1960). (a, b): Whittaker's representation of the variation in local abundance with moisture level, obtained by averaging the data and fitting normal curves. (Adapted from Whittaker 1967.) (c, d): The much more heterogeneous pattern of variation in local population density in these species as a function of moisture availability revealed by plotting Whittaker's original data. (Adapted from Brown 1984.)

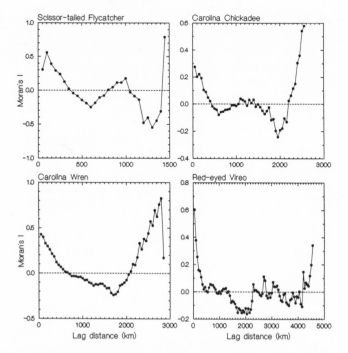

FIGURE 4.8. Spatial autocorrelation of abundance of four passerine bird species over their geographic ranges. The autocorrelation statistic, Moran's *I*, was computed from Breeding Bird Survey counts and is plotted as a function of the distance between census sites (lag distance, in increments of 50 km). Note the two peaks of high correlation, one for sites in close proximity (which can be anywhere within the range) and the other for sites at maximum lag distances (which must correspond to sites at opposite edges of the range). (From J. H. Brown, D. W. Mehlman, and G. C. Stevens, unpub.)

to, and to decline more precipitously at the edges of, ranges that coincide with major habitat discontinuances, such as coastlines, than at edges where no single obvious factor limits the range (D. W. Mehlman, unpub.). These are precisely the patterns predicted by the niche-based model.

Fourth, the geographic pattern of abundance is quite stable over time. The BBS has censused birds at the same sites for more than 25 years, and over this period the hot and cool spots tend to remain in the same places. I have more than a decade of censuses of annual plants from my experimental study area, and again, the distribution of relative abundance among sample plots has remained consistent. This relative constancy suggests that the characteristics of sample sites that determine local population density are fairly permanent features of the landscape. In the case of the birds, there have been some changes in abundance at certain census sites over two decades, but there have also been important changes in habitat: succession has occurred, forests have been cleared, and suburbia has encroached. We are examining whether known environmental changes can account for these changes in abundance.

Fifth, a simple computer model of a multidimensional Hutchinsonian niche superimposed on an environmentally heterogeneous landscape can produce patterns of variation in abundance similar to those described above. In constructing this model, we assumed that a hypothetical species is limited in abundance by the product of some small number of independent environmental variables or niche axes. The abundance factor contributed by each variable was determined by drawing from a truncated normal distribution, so that it could vary from zero to some maximum value. We have performed simulations using different numbers of niche dimensions, assuming that they were distributed among sites either at random or in some spatially explicit, autocorrelated pattern. The results resemble the patterns of abundance observed empirically, especially when the simulations used a relatively small number (2–6) of niche dimensions.

Evaluation: Alternative Hypotheses

We must be careful, however, because the niche-based model does not contain two potentially important influences on local abundance: dispersal and social density dependence. An obvious alternative explanation for all of the above data is that the organisms are behaving like soot particles and dispersing from some source area, rather than being distributed according to the suitability of local environments (see Grinnell 1922; Hanski 1982c; Holt 1983, 1992). Furthermore, Pulliam (1988) has shown that relatively simple models of dispersal among spatial patches of differing quality do not always give the intuitive result that equilibrium density is higher in the better patches. Sometimes populations in what Pulliam calls source habitats, presumably favorable locations where recruitment exceeds mortality and emigration exceeds immigration, can be less dense than those in lower-quality sink habitats, where death rates exceed birth rates and populations are sustained by immigration.

Our niche-based model implicitly assumes some density dependence. It assumes that individuals will distribute themselves over space in what Fretwell (1972; Fretwell and Lucas 1970) called an ideal free manner: that they pack into local areas so that their fitnesses in each habitat are equal, resulting in denser populations in the more favorable patches. Alternatively, as Fretwell suggested, social interactions can cause organisms to be distributed in a despotic way: some individuals are able to defend space or resources in the better habitats, resulting in lower densities in these favorable sites than would be expected on the basis of environmental quality and our niche-based model.

It is likely that both dispersal and social behavior affect the spatial distribution that we observe empirically. We know that individuals move, that their ability to assess environmental quality must be imperfect, and that many species of birds and other organisms exhibit territoriality or other forms of despotic behavior. These factors would tend to prevent them from being distributed in the ideal free way that the niche-based model assumes.[3] Whether the niche-based model is sufficient to account for the observed patterns, or whether any or all of these additional factors must be included, now becomes an empirical question. I suspect that social behaviors will have to be included in the model in order to accurately predict the local abundance of many species of birds and other organisms. Our computer simulations typically predict that a very small proportion of sites will contain as much as an order of magnitude more individuals than we observe empirically, even for the commonest birds. Territoriality or some other form of intraspecific repulsion is an obvious explanation.

While detailed behavioral and population studies are useful for documenting both despotism and dispersal, the roles of these processes in determining the spatial distribution of abundance may be better assessed using macroecological approaches. Doug Morris (in press) has developed an elegant regression technique to detect despotic distributions, and he finds that at high densities the white-footed mouse (*Peromyscus leucopus*) is distributed among different types of habitat in a despotic way.

Dispersal and source-sink population dynamics are very difficult to measure accurately in microscopic studies, because to do so usually requires that all individuals be marked and their fates followed; all immigrants must be identified and all individuals that disappear must be correctly identified as having either died or dispersed. The macroscopic approach offers an alternative. If the niche variables that limit local abundance can be identified (by correlating local abundance with environmental variables), then an effect of dispersal from favorable sources to less favorable sinks can be predicted: unfavorable sites (cool spots) near hot spots should have higher densities than equally unfavorable sites far from dense populations. George Stevens (pers. comm.) has some preliminary results that support this prediction. He has used discriminant analysis to determine the combinations of abiotic environmental variables for localities in Vermont where particular bird species were present and absent. He finds that the analysis tends to make incorrect assignments, predicting that a species will be present where the conditions are unsuitable at locali-

---

3. Social behavior can also cause organisms to be more clumped than would be expected if they were distributed in an ideal free way. An obvious example is any species that breeds in colonies or travels in flocks, herds, or schools.

ties near the center of the range, and predicting that it will be absent where conditions are favorable at localities at the edge of the range. This result is consistent with the hypothesis that dispersal and the presence of nearby populations affects the tendency of individuals to occur in marginally suitable environments.

Implications

The relationship between abundance and distribution is a prime example of the statistical approach of macroecology, applied in this case to the distribution of individual organisms of the same species over geographic space. Even the relatively crude data that are presently available permit us to extract general, emergent statistical patterns from the complex details of the actual distributions. And once we have these patterns we are challenged to begin developing and evaluating mechanistic hypotheses to account for them. At a minimum, this exercise has caused us to ask new and more sophisticated questions.

In particular, the spatial distribution of abundance raises questions about the connections between the micro and the macro, between the structure and dynamics of populations studied by ecologists and the distributions of species studied by biogeographers and systematists. On the one hand, a large-scale perspective emphasizes the need to question ecological models that assume local regulation of closed populations, and the need to study dispersal and its consequences. On the other hand, the small-scale perspective emphasizes the power of the niche concept to show how species-specific requirements and spatial variation in environmental conditions, which most immediately affect the survival, reproduction, and movement of individual organisms, ultimately determine the boundaries of the geographic range and the distribution of abundance within these boundaries.

THE RELATIONSHIP BETWEEN ABUNDANCE AND DISTRIBUTION AMONG SPECIES

Pattern

In chapter 2 of *The Origin of Species,* Darwin (1859) noted that within genera containing many species, there were often what he called "dominant species,-those which range widely over the world, are the most diffused in their own country, and are the most numerous in individuals," and other species that had the opposite combination of traits: small geographic ranges, restricted habitat distribution, and low abundances. More recently Rabinowitz (1981; Rabinowitz, Cairns, and Dillon 1986), Hanski

(1982c), and others (Bock and Ricklefs 1983; Bock 1984a,b, 1987; Brown 1984; Gotelli and Simberloff 1987; Morse, Stock, and Lawton 1988; Burgman 1989; Gaston and Lawton 1989, 1990a,b; Collins and Glenn 1990; Lawton 1990) have explored the relationship between these characteristics of species.

Most of these investigators have found the pattern that was reported

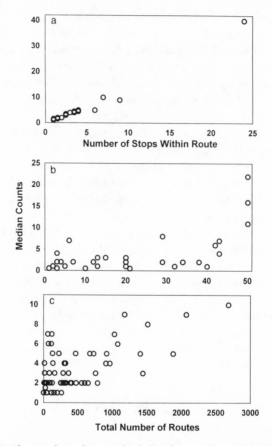

FIGURE 4.9. The correlation between local abundance and spatial distribution of the species of wood warblers (subfamily Parulinae) at three scales, based on census count data from the North American Breeding Bird Survey. (*a*) Local scale: number of individuals counted as a function of number of stops (out of 50, 0.8 km apart) where the species occurred within a single census route in the state of Maryland. (*b*) Intermediate scale: median number of individuals counted per census route as a function of the number of routes in the state of Maryland where that species was recorded. (*c*) Continental scale: median number of individuals counted per census route as a function of the total number of routes (out of approximately 1,800 dispersed across the United States and Canada) where the species was recorded. (Data analyzed by R. Pierotti and J. H. Brown [unpub.].)

by Darwin (e.g., fig. 4.9). When enough sites have been sampled so as to provide a relatively accurate measure of the average population densities of several closely related or ecologically similar species,[4] there is a positive correlation between local abundance and several measures of larger-scale spatial distribution: the abundant species tend to be widely distributed and the rare ones tend to have restricted ranges. Several things are noteworthy about this pattern:

1. It has been documented by many different investigators, working independently with organisms as different as plants, insects, and vertebrates (for a comprehensive survey of these studies, see Gaston and Lawton 1990a). This suggests that the pattern is very general.

2. The pattern is usually clear only when the analysis is restricted to species that are either closely related phylogenetically, such as those in the same genus as mentioned by Darwin, or ecologically similar, such as the members of the same ecological guild, functional group, or life form (which may be, but are not necessarily, also taxonomically related; see Root 1967; Wiens 1989; Simberloff and Dayan 1991). This suggests that the pattern in some way reflects the constraints of ecological relationships, phylogenetic history, or both.

3. As indicated by the quote from Darwin, there are positive correlations between different measures, made at different spatial scales, of the spatial distribution of species: within local habitat patches (as reflected in population density or abundance), among localities or habitat types within a region, and among regions (as reflected in the area of the geographic range). This, together with the previous observation, suggests that relatively small differences among otherwise similar species have pervasive effects on their ecological relationships, and that these effects are reflected in their spatial distributions.

4. The pattern is the opposite of what might be expected from certain kinds of trade-offs between specialists and generalists. At least in one sense, species with narrow habitat distributions and restricted geographic ranges might be considered to be specialists, whereas those with broad habitat and geographic ranges could be considered generalists. But the narrowly restricted species are not usually so well adapted to the few places where they occur that they are able to attain high densities there. On the contrary, even in the environments where they do occur, the narrowly distributed species are typically outnumbered by more widely distributed relatives and guild members. In this case, the jack-of-all-trades appears also to be a master of all.

---

4. Two methodological points are relevant here. First, in performing such correlation analyses it is important to use only sites where a species occurs to calculate its average abundance (i.e., to exclude the zero values). Second, because the frequency distribution of abundance within each species tends to be extremely right-skewed (fig. 4.5), the median or the geometric mean provides a much more accurate representation of the central tendency than the arithmetic mean.

Process

So how do we account for the positive correlations between abundance and distribution, and for the other features of these relationships noted above? I suggest a simple niche-based explanation. Consider two hypothetical species. One has broad requirements: it can tolerate a wide range of physical conditions, use many different kinds of resources, and survive in the presence of many potential enemies. Call this species broad-niched or a generalist if you will. The second species has the opposite traits: tolerance for only limited abiotic conditions, ability to use only a few kinds of resources, and high sensitivity to several competitors, predators, parasites, and diseases. This species could be called narrow-niched or a specialist. Now what would we predict about the abundances and distributions of these two species? The same attributes that would enable the first species to occur in many different habitats over a wide geographic area would also enable it to attain relatively high population densities in many of those places. In contrast, the narrow requirements of the second species would cause it to be restricted to a few habitat types and within a limited geographic area, and these same limitations of abiotic and resource requirements and susceptibility to biological enemies would prevent it from being abundant even in those places. This is the pattern Darwin described; the first species is what he called dominant.

More formally, the niche-based model for spatial variation in abundance within species can be extended to account for the pattern among species. Assume, as above, that the abundance and distribution of any given species reflects the extent to which individuals in local populations are able to satisfy their requirements in a spatially variable environment. It is necessary to make two additional assumptions. The first is tautological and therefore requires no justification: species differ in their requirements for abiotic and biotic conditions.

The second assumption is that closely related or ecologically similar species are constrained in the ways and the extent to which they differ from each other. Phylogenetic relatives are constrained by their common ancestry. New traits that arise in a species and distinguish it from its relatives must be derived from and functionally integrated with preexisting characteristics. This constraint limits the extent to which relatives differ. Somewhat similarly, species in the same ecological guild are limited in the extent to which they differ by being constrained to occur in similar environments and to use similar resources (Root 1967; Simberloff and Dayan 1991). Because of the multiplicative effect of niche variables, even a small difference on just one niche axis will tend to have substantial effects on abundance and distribution. For example, compared with an otherwise similar species with a narrower niche, a species that is slightly more tolerant of some abiotic condition or biotic interaction or is slightly better able to use some resource should not only be able to occur in more places but also to attain higher abundances in some of those places.

Furthermore, such differences in niche breadth will tend to be reinforced by a kind of positive feedback in adaptive evolution (Holt 1987; Holt and Gaines 1992). Species that already have broad niches will be selected to maintain and sometimes to increase breadth. Imagine a species whose northern limit of range is set largely by its tolerance for cold. Populations along the northern boundary are effectively continually under selection for increased cold tolerance, because this would allow the individuals to better survive and the populations to expand. Although selection might be less strong elsewhere in the geographic range, increased cold tolerance should also benefit individuals far from the northern range boundary, for example, by improving their performance during the winter.

Now consider a narrow-niched species, for example, a specialized parasite that can utilize only one host species with a small geographic range. The parasite's distribution will necessarily be limited to localities where its host occurs. Even though it might once have been broadly tolerant of other environmental conditions (perhaps because its ancestor infected a widely distributed host), there will be little or no selection to maintain tolerance for any conditions beyond those that it now experiences within the limited range of its current host.

My niche-based explanation for the positive correlation between abundance and distribution in similar species has the same limitations as the single-species model: it is completely deterministic and it contains no dynamics, and hence no time lags, dispersal, or despotic behaviors. Hanski (1982c, 1991; see also Hanski and Gyllenberg 1991; Holt 1983, 1992) has developed models showing that dispersal and the resulting stochastic metapopulation or colonization/extinction dynamics can contribute to and perhaps even cause the pattern. Hanski's models and empirical papers call attention to what he calls the "core-satellite" pattern: a bimodal frequency distribution of species with respect to the number of sample localities where they occur (fig. 4.10a,b). Hanski refers to the abundant species that are also found at the majority of sites as "core species," and to the rare species that are also restricted to a small proportion of sites as "satellite species." The essence of Hanski's models is that since the core species are both abundant within sites and distributed among many sites, they have a high probability of producing emigrants both to sites where they are

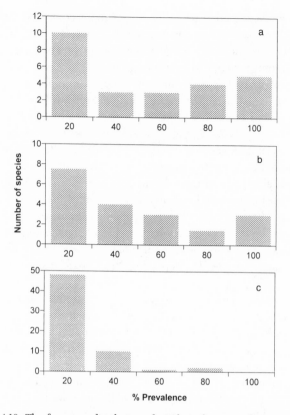

FIGURE 4.10. The frequency distribution of number of species of parasitic helminths as a function of the proportion of individual hosts (the willet, *Catoptrophorus semipalmatus*, a shorebird) infected at two spatial scales: *(a)* local scale, for a single locality in Alberta; *(b)* local scale, obtained by averaging data from nine local sites in Alberta, Manitoba, California, Louisiana, and Florida; *(c)* geographic scale, obtained by combining data for 61 species distributed among nine sites in Alberta, Manitoba, California, Louisiana, and Florida. Note that the bimodal "core-satellite" pattern emphasized by Hanski (1982c) is observed at the local scale, but when the same kind of data are collected on a much larger spatial scale, a clear unimodal pattern is obtained. Compare with the results of the computer simulations in figure 4.11. (Data from A. Bush; analyses by J. Hnida [unpub.].)

already present (and thus increasing the abundance) and to sites where they are at least temporarily absent (thus colonizing new sites or recolonizing ones where local extinction has occurred). Similarly, the satellite species, because of their initially low densities and restricted distribution, will tend to produce few successful dispersers, thus maintaining the status quo.

Hanski's metapopulation dynamics models and my niche-based model appear to offer alternative explanations for the positive relationship between abundance and distribution. However, the processes invoked in these two kinds of models are by no means mutually exclusive. In fact, it is likely that the probabilities of colonization and extinction depend, at least in part, on the niche requirements of species. For example, narrow-niched species with small local populations and geographically restricted, regionally patchy distributions would be expected to have high probabilities of local extinction. Conversely, the extent to which local abundance reflects tolerance for local environmental conditions will depend, at a minimum, on the species' having sufficient dispersal capability to colonize suitable sites and to recolonize them after local extinction.

Even though they are not mutually exclusive, the metapopulation dynamics and niche-based models often suggest different, testable hypotheses to account for empirical patterns. For example, the niche-based model would suggest a positive correlation between abundance and distribution of species even among islands of archipelagoes where over-water dispersal is too low have any appreciable effect on population density. While I do not know of any direct test of this prediction, I believe that an analysis of Caribbean land birds would support it. The most abundant species on each island tend to be widespread species rather than single-island endemics (pers. obs., and see Ricklefs and Cox 1972; Cox and Ricklefs 1977).

Metapopulation and niche-based models also suggest alternative explanations for the core-satellite bimodal frequency distribution of species among local or regional sample sites. Hanski's models assume that the sites are all identical and hence equally suitable for habitation by all species. The differences among species in the proportion of sites occupied thus reflect dispersal capabilities and the stochasticity of colonization/extinction dynamics. Alternatively, the niche-based model suggests that sites might be very similar but never exactly identical. Thus small differences in the requirements of species could produce the core-satellite pattern: some of the species, those with broad niches, would be able to inhabit all of the sites, whereas other species, those with narrow requirements, would be able to occur in only the few sites that met their more stringent requirements.

Pablo Marquet and I (unpub.) tested the feasibility of the latter explanation by computer simulation. We distributed niches of species at random along a single niche axis. Then we sampled at "sites" along this axis and plotted frequency distributions of species as a function of number of sites inhabited. When we sampled sites representing a narrow range of environmental variation, we obtained a bimodal distribution similar to the core-satellite pattern (fig. 4.11b; compare with figure 4.10a,b). When we

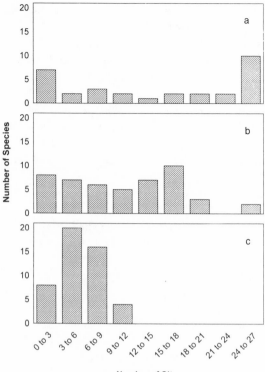

FIGURE 4.11. Results of computer simulations by Pablo Marquet and myself showing how a simple model based on niche requirements predicts a bimodal "core-satellite" frequency distribution of species among sites at a local scale and unimodal distributions at larger scales. The model assumed that niches of "species" were distributed at random along a single niche axis. *a, b,* and *c* represent the results of three simulations in which the mean and variance in niche breadth of the species were held constant, but "sites" were sampled from an increasing range of environmental variation along the niche axis. As the samples were taken over a wider range of positions on the axis, corresponding to more variable and/or more spatially separated sites, the frequency distribution changed shape, from distinctly bimodal (*a*) to unequivocally unimodal (*c*).

sampled sites representing a wider range of variation along the niche axis, we obtained a unimodal distribution (fig. 4.11c). A similar pattern is observed in field data as the spatial scale of sampling varies from a local area to a large biogeographic region (fig. 4.10; see also Brown 1984, his fig. 9). This suggests that the bimodal distributions that are observed empirically, even when species are sampled from similar sites in close proximity, may reflect niche requirements more than metapopulation dynamics. Of course, both processes are almost certainly at work at local scales, so their

TABLE 4.1. Distribution of 160 species of British plants with respect to pairwise categories of geographic range, habitat specificity, and local population size.

Geographic distribution	Wide		Narrow	
Habitat specificity	Broad	Restricted	Broad	Restricted
Local population size — Somewhere large	58	71	6	14
Local population size — Everywhere small	2	6	0	3

Source: Rabinowitz, Cairns, and Dillon 1986.

relative importance becomes an empirical issue that can only be addressed by appropriate observations and experiments in the field.

Exceptions

There often seem to be specific exceptions to statistical macroecological patterns. In the case of interspecific patterns of abundance and distribution, there are good examples of such exceptions. Rabinowitz and her co-workers (Rabinowitz 1981; Rabinowitz, Cairns, and Dillon 1986), for example, used essentially the same attributes as Darwin (i.e., local abundance, distribution among habitats, and geographic range) to categorize British plants (table 4.1). They found examples of species that exhibited seven of the eight combinations of traits in their qualitative classification, rather than just the two that Darwin might have expected. Three things are noteworthy about these results. First, Rabinowitz et al. did not restrict their analyses to species that were either closely related taxonomically or particularly similar ecologically. Second, even in this heterogeneous assemblage of species, they could find no species in one category (wide geographic range, broad habitat specificity, and small local populations). The overwhelming majority of species (129 out of 160) were in two categories (wide geographic range and locally abundant, with either broad or restricted habitat requirements). Third, as interesting as these findings are, they are a bit hard to interpret because all species were put into one of two alternative classes for each attribute, even though many of them may have exhibited intermediate characteristics.

More quantitative analyses also reveal exceptions, however. Ray Pierotti and I (unpub.) have identified several examples of closely related and/or ecologically similar bird species that are usually very abundant despite their restricted habitat and geographic ranges (e.g., Lucy's warbler and tricolored blackbird).[5] These appear to be species that have requirements for narrow, but highly productive, environmental conditions.Pierotti and I did not find any unequivocal examples of species that are locally rare but widely distributed, so long as comparisons were restricted to closely related or ecologically similar forms.

Implications

Despite the century that has passed since Darwin's observation, the implications of relationships between abundance and distribution are just beginning to be appreciated. The fact that the patterns tend to hold only for assemblages of phylogenetically related or ecologically similar species suggests that both evolutionary and ecological constraints severely limit the combinations of traits that species, and the individual organisms within these species, can possess.

The evolutionary constraints emphasize the importance of phylogenetic studies in ecology. The great advances in phylogenetic systematics during the last few decades are just beginning to be appreciated in ecology. Thus, for example, in nearly all of the studies cited above, comparisons of "closely related species" were made by restricting the analysis to species of some equivalent taxonomic rank, such as the same genus or family. But these species differ considerably in their actual relatedness to one another. For one thing, this method causes statistical problems, because sister species should not be treated as independent samples (e.g., Felsenstein 1985; Pagel and Harvey 1988, 1989; Harvey and Pagel 1991; Gittelman and Luh 1992). The development of accurate and practical methods for phylogenetic reconstruction and statistical analysis should permit much more precise and rigorous investigations of the ecological, biogeographic, and macroevolutionary consequences of the constraints owing to evolutionary history. The few such studies done thus far (e.g., Gittelman 1985; Pagel and Harvey 1988, 1989; Brooks and McLennan 1991; Harvey and Pagel 1991; Taylor and Gottelli, in press) suggest that they will be extremely informative.

Phylogenetic reconstructions give us a rigorous way to measure the extent to which species are "closely related." We do not have a comparable way to quantify the extent to which species are "ecologically similar," which limits our ability to investigate further the causes and consequences of ecological constraints. The fact that guilds of species that are not neces-

5. I suspect that many species endemic to isolated islands are also as abundant locally as closely related or ecologically similar species that have much larger ranges on continents. This reinforces the point, made in chapter 3, that continents may not provide a good model for islands, and vice versa.

sarily closely related also exhibit predictable relationships between abundance and distribution suggests that there are a limited number of biological solutions to environmental challenges. This is the premise that underlies studies of evolutionary convergence (see discussion and references in chapter 3).

Much of the variation among closely related and/or ecologically similar species is in what Darwin called dominance. Local abundance is one, but only one, expression of dominance. Darwin clearly suggests that species vary in overall quality or fitness.[6] The degree of such dominance depends on traits of species that interface with characteristics of the abiotic and biotic environment to affect survival and reproduction. Furthermore, the combinations of traits possessed by different species generally do not result in a trade-off between specialists and generalists. Instead, species appear to be arrayed largely along a spectrum of dominance. This fact has obvious important implications for macroevolution and conservation.

OTHER CORRELATES OF ABUNDANCE AND DISTRIBUTION

Predictions

It should be possible to make predictions about characteristics of individual organisms from information about the abundance and distribution of species. Two kinds of predictions are possible. First, individuals of abundant, widespread species should have broader tolerances and requirements for both abiotic and biotic conditions than individuals of rare, restricted species. This may seem obvious, because the abundant, widespread species certainly cope with a much wider range of conditions over their geographic ranges. But it is important to test this prediction for two reasons.

First, it is important to know whether species-level traits that affect ecological relationships such as abundance and distribution can indeed be associated with characteristics of the morphology, physiology, and behavior of individual organisms. The answer to this question says much about the extent to which we can hope to develop lower-level, reductionist explanations for macroecological phenomena. Unless this can be done, we have little prospect for understanding the underlying mechanisms.

---

6. It is often suggested that Darwin advocated a very gradualistic view of evolution, based on the differential survival and reproduction of individual organisms with slightly different characteristics. A careful rereading of *The Origin of Species,* especially chapter 2, should set the record straight. Darwin was concerned not only with what we now call microevolution, but also with macroevolution: with the origin, survival, and extinction of species.

Second, and perhaps even more importantly, it is certainly not obvious that broad-niched species must be composed of broad-niched individuals. The variation in requirements that enables a species to attain a broad geographic distribution or even a high local population density might in theory depend entirely on variation among, rather than within, individuals. Thus, geographic variation among local populations of widespread species and polymorphic variation among individuals within local populations might account for some or all of the variation in niche breadth. The interpretation and implications of the positive correlation between abundance and distribution depend on the extent to which the variation occurs within or among individuals of a species (e.g., see Roughgarden 1979).

Given the importance of addressing both of these issues, it is perhaps surprising that there have been relatively few attempts to compare the requirements of closely related or ecologically similar species that differ conspicuously in abundance or distribution. However, the few studies that have been done permit some preliminary conclusions.

A significant component of the niche breadth of most widely distributed species probably can be attributed to geographic variation; that is, the species are composed of differentiated populations of individuals that are somewhat adapted to the local environment. Darwin noted that what he called dominant species tend to be differentiated into regional varieties. Many subsequent studies have shown that at least some of this geographic variation is adaptive, facilitating the survival and reproduction of individuals in the local environment, and—at least by inference—enabling the species as a whole to occupy a wider geographic range. For example, classic early studies of the deer mouse (*Peromyscus maniculatus*) by Sumner (1932) and Dice (1947; also Dice and Blossom 1937) showed that the pelage color of mice tended to match the surrounding environment and that this was advantageous in avoiding capture by predators. Horner (1954; see also Thompson 1990) demonstrated differences among deer mouse populations in morphology and behavior that affect climbing ability and reflect adaptations to open (desert and grassland) or more vegetated (woodland and forest) environments. Hayward (1965a,b) failed to find evidence for significant differentiation in the thermoregulatory physiology of local deer mouse populations living in regions with different climates, but other studies of small mammals have demonstrated such adaptive variation (e.g., Brown 1968).

Several studies have identified variation in morphological and physiological traits among individuals within populations that are correlated with differences in growth, survival, and reproduction (e.g., Ehleringer 1984a,b; Dunham, Grant, and Overall 1989). It is much more difficult, however, to demonstrate that this variation per se is adaptive and allows the population to exploit a wider range of resources and microhabitats. There is evidence, although much of it is circumstantial, that within-population polymorphism is adaptive because it increases niche breadth. The classic example is the huia (*Neomorpha acutirostris*), a now extinct trunk-foraging New Zealand bird in which the two sexes had bills of completely different shapes, apparently enabling pairs to exploit a much wider range of food resources than could be used by an individual of either sex. In mammalian carnivores of the genus *Mustela* (weasels) and hawks of the genera *Accipiter* and *Falco* there is dramatic sexual size dimorphism, with males and females of the same population often being as different in size as members of different sympatric species (see Schoener 1984; Dayan, Simberloff et al. 1989). These differences enable the sexes to use different prey.

Notwithstanding the important contribution of adaptive variation among individuals within and between populations, there is also evidence that single individuals of dominant species tend to have broader tolerances and requirements than individuals of rare, restricted species. For example, populations of pupfish (genus *Cyprinodon*) species that inhabit a wide range of thermal environments are composed of broadly tolerant individuals. Members of populations that have been relatively recently isolated in thermal springs of constant temperature have retained the broad tolerances of their widely distributed ancestors (Brown and Feldmeth 1971; Feldmeth 1981). Members of endemic species that have a longer history of isolation in springs of extremely high and constant temperature, however, are unable to tolerate such low or varying temperatures as individuals of closely related, widely distributed species that inhabit thermally fluctuating environments (Feldmeth 1981; D. L. Soltz and C. R. Feldmeth, pers. comm.). Since the stenothermal, narrowly endemic species are descended from eurythermal, widely distributed ancestors, this also supports the suggestion made above that niche breadth tends to decrease in the absence of continual selection to maintain broad requirements.

At the moment there is probably not sufficient information to assess the relative contributions of the within-individual and between-individual components to overall niche breadth and to the relationship between local abundance and larger-scale distribution described above. Both almost certainly play significant roles, and they may reinforce each other. The situation is further complicated by the fact that not all of the apparent differences among individuals and populations need be genetic; a substantial component may be due to the phenotypic plasticity of broad-niched individuals (e.g., thermal tolerances of pupfish; see Brown and Feldmeth 1971; Feldmeth 1981).

The second kind of prediction about characteristics of individuals that can be made from the abundance and distribution relationships of species concerns what are called "fitness traits," characteristics that contribute directly to an individual's survival and reproduction. On the one hand, these kinds of traits are thought to have relatively little heritable variation within or among species. Since, everything else being equal, it is always better to reproduce at a younger age, have larger litters, reduce the interval between litters, and prolong the reproductive life span, it is argued that there is always strong directional selection tending to use up the variability in these kinds of traits. (Let me note parenthetically, however, that the same argument can be made for any trait that increases niche breadth: if everything else really is equal, it will also always be advantageous to be tolerant of a wider range of temperatures, protectively colored, able to use a greater variety of food types, and so on). On the other hand, if some species really are dominant over their relatives, as Darwin suggested, then we would predict that this would be reflected in interspecific variation in fitness traits. Thus we might expect abundant, widely distributed species to have larger litters, younger ages of reproduction, and so on.

Although I am aware of only a few relevant studies, their results appear to support this prediction. Glazier (1980) found a significant positive correlation between area of geographic range and litter size among species of mice in the genus *Peromyscus*. Subsequently, my former student Alan Harvey (pers. comm.) tested for a similar relationship between clutch size and area of geographic range in several large genera of North American and Australian birds. He, too, found significant positive correlations, with two exceptions: hole nesters showed no relationship, and island endemics had higher clutch sizes than would be expected from their small geographic ranges. I find these preliminary results encouraging, especially since hole-nesting birds are known not to conform to other patterns of clutch size variation and island populations often tend to have exceptionally high population densities. This is an area in which there is clearly opportunity for much additional work. So far, most of the analyses have used the readily available data on geographic range and litter size, but it would be interesting to analyze other kinds of fitness traits and other measures of abundance and distribution. If many kinds of traits of individual organisms can be correlated with simple measures of abundance and distribution, this should provide the basis for investigating important connections between patterns that are revealed at the level of populations, entire species, and clades of related species and mechanistic processes that operate at the level of the morphology, physiology, and behavior of individual organisms. This is just one of many cases in which the macroecological approach appears to offer new and valuable insights into more microscopic phenomena.

CONCLUDING REMARKS

To a large extent, abundance and distribution reflect the consequences of the same ecological relationships: the ways that the niche requirements of species and the dynamics of populations interact with spatial and temporal variation in the environment. It is perhaps not surprising, therefore, that there are usually positive correlations between local abundance and extent of spatial distribution, both within species and among closely related and/or ecologically similar species. Darwin was perhaps the first to call attention to the pattern of correlated suites of traits, including abundance, habitat and geographic distribution, and differentiation into regional varieties or races, among related species such that they form a hierarchy of "dominance."

Investigation of the mechanisms underlying these patterns promises to contribute importantly to our understanding of the relationships between different levels of organization and between different spatial and temporal scales: between the morphological, physiological, and behavioral attributes of individual organisms and the abundance and distribution of populations and species and between the local, short-term processes of organism-environment interaction, population regulation, and interspecific relationships traditionally studied by ecologists and the regional to global phenomena of geographic range limitation and colonization/speciation/extinction dynamics studied by biogeographers and macroevolutionists.

5 The Composition of Biotas

Patterns of Body Size, Abundance, and Energetics

I now turn from considering how species are distributed in space to focus on the characteristics of the species that occur together on different spatial scales. In a sense I am returning to one theme of chapter 3, community structure and assembly rules, but the approach will differ in two respects. First, rather than emphasizing the differences among the few species in the same guild that affect their coexistence, I will use the overall statistical distributions of parameter values for all species in much larger taxonomic groups to characterize the organization of these species-rich assemblages. Second, rather than focusing exclusively on those species that coexist within local communities, I will analyze the composition of assemblages of species that occur together over a wide range of scales, from local communities in small patches of habitat to entire continents.

BODY SIZE

I begin by examining the distribution of body sizes among species. This is a good starting point for two reasons. First, body size is easy to measure—all it requires is a ruler (or pair of calipers) to measure linear dimensions or a scale to measure mass. Consequently, lots of data are available, and there is a long history of studies of variation in body size within and among species. Second, many morphological, physiological, behavioral, and ecological traits are closely correlated with body size (e.g., McMahon 1973; McMahon and Bonner 1983; Peters 1983; Calder 1984; Schmidt-Nielsen 1984; La Barbera 1986; Bonner 1988). These relationships tend to be approximately linear when plotted on logarithmic axes. They are fitted with power functions, called allometric equations, of the form

$$V = CM^z, \tag{5.1}$$

where V is some measured variable, C is a constant, M is individual body mass, and z is another constant (the slope when plotted on logarithmic axes) that often takes certain specific values. Such allometric relationships reflect integration of the phenotype and the role of structural and functional constraints. I will have much more to say about them.

Size Frequency Distributions

In 1959 Hutchinson and MacArthur plotted the frequency distributions of body sizes among the species of land mammals of Michigan and Europe (fig. 5.1). They noted that these distributions were highly skewed, such that there were many more species of relatively small mammals than of large or extremely small ones. They suggested that this pattern reflected the capacity of the modal-sized species to be relatively more specialized, and hence to subdivide space and resources more finely. They suggested that the environment could be visualized as comprising a number of "mosaic elements." Thus the niche of each species would include a different combination of elements, with modal-sized species requiring smaller numbers of elements than larger or smaller species. More recently this hypothesis has been recast in more explicit terms using fractal geometry (May 1986, 1988; Morse, Stock, and Lawton 1988; Lawton 1990).

Since Hutchinson and MacArthur's study, several investigators have analyzed the frequency distributions of body sizes among species for dif-

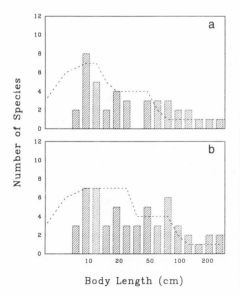

FIGURE 5.1. Hutchinson and MacArthur's plot of the frequency distribution of number of species as a function of body length (histogram bars) for the mammals of (a) Michigan and (b) Europe, also showing the distribution of values predicted by their model (dashed lines). Note that the abscissa is scaled logarithmically, and that their model correctly predicts a right-skewed distribution, but incorrectly predicts too many species of small size. (Adapted from Hutchinson and MacArthur 1959.)

ferent kinds of organisms (Van Valen 1973a; May 1978, 1986, 1988; Bonner 1988; Dial and Marzluff 1988; Morse, Stock, and Lawton 1988; Lawton 1990; Brown and Nicoletto 1991; Maurer, Brown, and Rusler 1992). Whenever they have considered a diverse assemblage of species from a large region, they have almost invariably obtained very similar results: the frequency distributions are highly skewed, even when body size is plotted on logarithmic axes (fig. 5.2). Note that when logarithmic axes are used, the same pattern is found regardless of whether data on body length (L) or body mass (M) are used (compare figs. 5.1 and 5.2). Because of the allometric relationships mentioned above, mass is very closely correlated with length.

Groups of organisms as different as bacteria, trees, insects, fishes, and mammals all show the pattern described by Hutchinson and MacArthur: many more species of small size than of large. May (1978) made a heroic

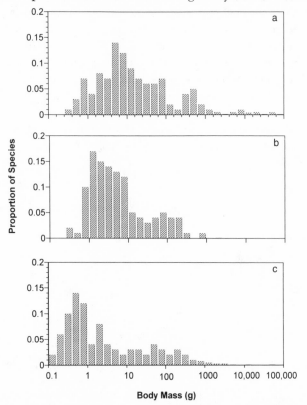

FIGURE 5.2. Frequency distributions of numbers of species as a function of body mass (on a logarithmic scale) for three groups of vertebrates in North America: (*a*) terrestrial mammals; (*b*) land birds; (*c*) freshwater fishes. Note that all three distributions have essentially the same right-skewed shape. (Adapted from Brown, Marquet, and Taper 1993.)

attempt to estimate the diversity of all organisms as a function of their size, and found the same pattern. A relationship that is apparently so general begs for explanation. Is Hutchinson and MacArthur's ecological specialization explanation sufficient to account for these patterns? That is a difficult question to answer. Although Hutchinson and MacArthur's paper was insightful, it is hard to know just what they mean by "mosaic element" and "niche."[1] Rather than attempting an answer now, I will first consider some of the data and their implications in more detail.

First, note that on a logarithmic scale the data in figure 5.2 are not normally distributed (i.e., they are not symmetrical, lognormal distributions), but instead highly right-skewed. This seemingly esoteric point is important, because just as normal distributions are produced by additive combinations of random variables, lognormal distributions are produced by multiplicative combinations of random variables (May 1975). The sizes of many physical objects, from the grains of sand on a beach to the heavenly bodies in the Milky Way, are lognormally distributed, presumably reflecting the multiplicative effects of stochastic processes. That body sizes deviate significantly from lognormal distributions makes them more interesting and suggests that it might be worthwhile to search for deterministic mechanisms that produce distributions of this special form.

There are several things to explain:

1. The distribution of body sizes is highly modal: there is some particular logarithmic size category that contains the largest number of species.

2. From the mode there is a gradual decline in numbers of species of progressively larger size, producing the long tail and pronounced right skew of the distribution.

3. From the mode there is a sharp decline in numbers of species of progressively smaller size. Note that this last characteristic already points out a limitation of the specialization hypothesis of Hutchinson and MacArthur (1959; see also Morse, Stock, and Lawton 1988). If specialization promotes diversity and is inversely related to body size, why isn't the greatest species diversity in the smallest size categories?

Insights into the processes contributing to the pattern might come from considering variation in the shapes of the distributions. Brown and Nicoletto (1991) found consistent variation in the form of the frequency distributions for North American mammals with spatial scale. The distribution for all 465 species in the continental fauna had the same right-skewed form described by Hutchinson and MacArthur and also found in other taxonomic groups (fig. 5.2a). When samples of species from smaller

1. Furthermore, others (e.g., Eisenberg 1981) have made just the opposite conjecture: that modal-sized species are the most generalized, and able to use the widest range of resources.

areas within North America were analyzed, however, the shape of the distributions changed. Samples of local communities, those species that coexist in small patches of nearly uniform habitat, tended to have uniform distributions, with nearly the same range of sizes represented but an approximately equal number of species in each logarithmic size category (fig. 5.3). And samples of biomes, areas of intermediate size containing generally similar climate and vegetation, had intermediate shapes.

Since the samples of small areas were nested within the samples of larger ones, the flattening of the distribution with decreasing spatial scale must be due to the more rapid turnover in space of modal-sized as compared with larger or smaller species; that is, the species of the modal size tend to differ among the smallest-scale samples, while the species of large and extremely small size tend to occur repeatedly in different samples. Thus, on average, the modal-sized species tend to occupy narrower ranges of habitat and/or smaller geographic ranges than species of extreme sizes. Roger Harris (pers. comm.), analyzing Lawton's data for insects on bracken fern in Britain, has found the same tendency for the frequency distribution of sizes to become flatter at smaller spatial scales. This finding is certainly consistent with Hutchinson and MacArthur's (1959) suggestion that the modal-sized species—but not the smallest species—are in some ways more specialized than their larger or smaller relatives.

In a recent paper, Holling (1992) performs a somewhat similar analysis of body sizes of North American land mammal and bird species that occur together in different types of habitats. However, he calls attention to the distribution of sizes in clumps separated by gaps. After evaluating and rejecting several alternative hypotheses, Holling suggests that "a small set of plant, animal, and abiotic processes structure ecosystems across scales in time and space," and that "animals living in specific landscapes . . . demonstrate the existence of this lumpy architecture by showing gaps in the distribution of their sizes." While I find the demonstration of the clumpy/gappy pattern convincing, I am less sure of Holling's explanation, which seems very difficult to test. As an alternative, I suggest a two-part explanation: (1) certain species or sets of closely related species have wide distributions, occurring repeatedly and occupying similar size categories in different habitats; and (2) these repeating species interact with the other species (e.g., through competition and character displacement), affecting the sizes that can invade and coexist in each habitat. Thus, a few widely distributed, closely related sets of species could produce the resonant frequencies of body sizes and the gaps between them that remain relatively unchanged over many habitat types and large areas of the continent.

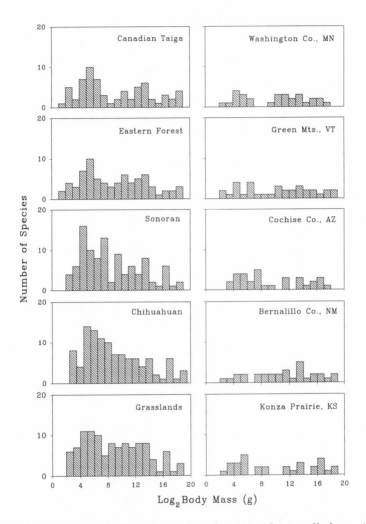

FIGURE 5.3. Frequency distributions of numbers of species as a function of body mass (on a logarithmic scale) for terrestrial North American mammals on two spatial scales: left, in biomes or large regions of generally similar vegetation; right, within small patches of relatively uniform habitat within the biome to the left. Note that both distributions are much flatter than the comparable pattern for the entire continent (fig. 5.2a), and that the local habitat patches have an approximately equal number of species in each logarithmic size category. (Adapted from Brown and Nicoletto 1991.)

Fractal Geometry and Ecological Specialization

May (1986, 1988), Morse, Stock, and Lawton (1988), and Lawton (1990) have made Hutchinson and MacArthur's hypothesis more explicit by casting the idea of specialization in terms of fractal geometry. The basic idea of fractals, developed by the mathematician Benoit Mandelbrot (1983; Feder 1988) is simple (fig. 5.4). Many physical entities exhibit "self-similar" structures over a wide range of spatial scales. Familiar examples are coastlines, which appear to be equally irregular regardless of the scale on which they are mapped, and snowflakes, which preserve similar geometric structure over a wide range of scales because of the hierarchical way in which water molecules are assembled to form ice crystals. The suggestion is that the environment has fractal properties such that it can hold many small, specialized species, each of which has highly specific niche requirements, but fewer large species, each of which is constrained to have broader requirements. For example, imagine species of progressively increasing size using successively larger triangles in figure 5.4.

There is probably much merit in this idea. In its present development, however, it has two shortcomings. First, it can account only for the decline in the number of species from the modal size to the largest size categories. Since it would predict that the smallest size categories in each taxon

FIGURE 5.4. The Sierpinski gasket, an example of a fractal pattern. It is easy to imagine how this might be taken to represent the packing of species of different body sizes (or the niches of those species), with there being more small species "inhabiting" the smaller triangles. (From La Brecque 1992.)

should be most specialized and therefore most numerous, some other hypothesis is needed to explain why this prediction does not hold. Second, the fractal hypothesis suffers some of the same problems in defining specialization that plagued Hutchinson and MacArthur. What does it mean to say that the environment has a fractal structure and that the breadth of niches is proportional to the body sizes of species? The often seemingly linear decrease in the number of species in logarithmic classes of increasing size has a slope (exponent of the allometric equation) in the range -2.0 to -3.0, suggesting a self-similar pattern with a fractal dimension of approximately 1.5 (May 1986; Morse, Stock, and Lawton 1988; see also Lawton 1990). It is not clear, however, just what characteristics of either environments or species' requirements one would measure to make this idea operational, so that it could be tested or form the basis of an empirical research program.

Physiological Constraints and Size-Abundance Relationships

A somewhat different kind of explanation for the Hutchinson-MacArthur pattern invokes physiological and ecological allometric constraints of body size. These arguments consider the effects of body size on the kinds and amounts of resources required to support individual organisms. Any resource available in the environment could seemingly support either many small organisms or a few large ones. Thus, if there is some relationship between the number of individuals and the number of species within a size class, the same resource should support more species of small body size than of large.

Before getting into the division of resources among species within size classes, however, it is necessary to point out that the capacity of resources to support individuals does not vary linearly with their body mass. Instead, the relationships are curvilinear power functions, because of two general features of physiological allometry. First, the requirements of an individual for energy or nutrients[2] vary as a fractional exponent of its body size: total requirements or metabolic rates per individual (E) scale as

$$E = C_1 M^{0.75}, \tag{5.2}$$

whereas mass-specific requirements or metabolic rates per unit mass (E/M) scale as

$$E/M = C_2 M^{-0.25}. \tag{5.3}$$

---

2. This equation describes the relationship between body size and rate of metabolism, and thus it characterizes the allometric scaling of requirements, not only for energy, but also for limiting nutrients (such as nitrogen for many herbivorous animals; Peters 1983).

Thus large organisms require more energy and nutrients per individual than small organisms, even though they require fewer resources per gram because their cellular metabolism is operating at a lower rate. These allometric relationships are very general. They apply to all organisms from bacteria to whales, although there is some variation in the value of the constants in the allometric equations fitted to data for different taxonomic groups (e.g., Peters 1983; Calder 1984).

The second ecologically relevant physiological consequence of body size is that the quality of food required by animals also varies allometrically. Equation (5.3) shows that the requirements of individuals for resources increase less rapidly with body size than the mass that is supported. Furthermore, because of the way that the anatomy and physiology of digestive systems scale with size, larger organisms have longer digestive tracts, retain food in the gut for longer times, and hence are able to extract a larger proportion of its energy and nutrients. As a result, large animals are able to meet their requirements by ingesting lower-quality foods than their smaller relatives can use (Townsend and Calow 1981; Peters 1983). Available data sets, many of them for large herbivorous mammals (e.g., see Owen-Smith 1988 and included references), suggest that the quality (Q) of food required (the reciprocal of the amount of energy and nutrients that can be extracted) scales as

$$Q = C_3 M^{-0.25}. \qquad (5.4)$$

Now we come to the thorny issue of how many individuals of different body sizes can be supported by a given area of appropriate habitat. For organisms, such as land plants and sessile intertidal invertebrates, for which space is the limiting resource, population density (N) might be expected to scale linearly and inversely with body mass:

$$N = C_4 M^{-1.0}, \qquad (5.5)$$

but this ignores many complications of how individuals of different sizes can be packed into two- or three-dimensional space.

For the majority of animals, which are more likely to be limited by energy and nutrients than by physical space, a first prediction might be that individuals of all sizes use nutritional resources at equal rates (Damuth 1981, 1987). Thus there should be a trade-off between body size and abundance, such that population density (N) should scale as

$$N = C_5 M^{-0.75}, \qquad (5.6)$$

the reciprocal of metabolic requirements. This prediction obviously ignores many potentially important considerations, such as the availability of resources of varying quality and the allocation of resources among species of both similar and different sizes.

Despite these problems, the empirical data offer some support for the above predictions. There is almost always a highly significant negative relationship between population density and body size, and often the slopes (exponents) of the allometric equations fitted to the data are between –0.75 and –1.0 (e.g., Damuth 1981, 1987, 1991; Peters 1983; Marquet, Navarrete, and Castilla 1990). However, there are potentially serious sampling biases in many of the data sets that have been analyzed. In particular, there may often be a tendency either to miss or to underrepresent very rare species (Brown and Maurer 1987; M. Silva and J. H. Brown, unpub.). More importantly, there is often so much variation in the abundances of species of similar size that the fitted regression equations account for only a small proportion of the observed variation. Having acknowledged that body size strongly influences abundance, let me defer further consideration of this topic for a few paragraphs.

All of the arguments about physiological constraints made above, like the arguments about ecological specialization, fail to consider the factors affecting the numbers of individuals and species in the smallest body size categories in each taxonomic group. The empirical data clearly suggest that not only are there fewer species in these smallest size categories (e.g., fig. 5.2), but these species also do not attain such high population densities as many of their larger relatives (see below and fig. 5.7).

This is not meant to imply, however, that the hypotheses about ecological specialization or physiological constraints should be discarded. Far from being mutually exclusive, the two hypotheses have much in common. The physiological constraints hypothesis begins to delve into the mechanisms that underlie the relationship between body size and specialization. Together, these hypotheses may go a long way toward explaining why the number of species in virtually all taxa decreases with increasing body mass from the modal size to the largest size category.

Energetic Constraints and Fitness

Recently, Pablo Marquet, Mark Taper, and I have developed a third hypothesis for the distributions of body sizes among species. It is based on a model of allometric energetic constraints on fitness (Brown, Marquet, and Taper 1993). We define fitness as the rate at which resources (energy, nutrients, and other essential commodities, such as water) in excess of those required for growth and maintenance of the individual can be harvested from the environment and utilized for reproduction. Thus, we equate fitness with reproductive power, the rate of conversion of energy into useful work for reproduction.[3]

Reproductive power is modeled as the consequence of two limiting rates: (1) the rate at which an individual can acquire resources from its environment and (2) the rate at which it can convert those resources into reproductive work. We assume that the process is analogous to a two-step chemical reaction in which the individual, I, catalyzes the conversion of resources, R, to work, W:

$$R + I_0 \overset{K_0}{\to} I_1 \qquad (5.7)$$

$$I_1 \overset{K_1}{\to} I_0 + W \qquad (5.8)$$

where I_0 and I_1 represent individuals before and after the acquisition of resources from the environment; they can be thought of either as the proportion of time a single individual spends in these two states or as the proportion of the two kinds of individuals in the population.

Reproductive power of an individual at steady state is given by

$$\frac{dW}{dt} = \frac{K_0 K_1}{K_0 + K_1} \qquad (5.9)$$

We further assume that the maximum values of both K_0 and K_1 that can be expressed by an individual are allometric functions of body mass, such that K_0 scales as M_0^b and K_1 as M_1^b. We assume that b_0, which scales the rate of energy acquisition in excess of maintenance needs, is 0.75, the same as the allometric exponent for individual metabolic rate, productivity, and growth rate (eq. [5.1]; Peters 1983; Calder 1984; Schmidt-Nielsen 1984). We assume that b_1, which scales the rate of transformation of energy to reproductive work, is −0.25, the same as the allometric exponent for the rate of mass-specific metabolism and nearly all biological conversion processes (eq. [5.2]; Peters 1983; Calder 1984; Schmidt-Nielsen 1984). Thus, we have

$$K_0 = C_6 M^b \qquad (5.10)$$

and

3. This concept of fitness as reproductive power deliberately ignores many of the complications of life history, especially the facts that an individual can allocate resources to either reproduction or survival, and that there is a trade-off between immediate reproduction and survival and future reproduction. As we have defined it, reproductive power includes the accumulation and allocation of resources for survival as well as reproduction. By harvesting and assimilating resources, individuals acquire energetic and material capital that can be invested both to survive and to produce offspring. Since all organisms are mortal, however, their fitness ultimately depends on the rate at which their capital is converted into reproduction.

$$K_1 = C_7 M^b. \tag{5.11}$$

Finally, maximizing reproductive power, dW/dt, with respect to body mass gives the optimal body mass, M°, as

$$M^\circ = \left(\frac{-C_1 b_0}{C_0 b_1}\right)^{\frac{1}{(b_0 - b_1)}} \tag{5.12}$$

The biological interpretation of this model is straightforward but informative. The smallest individuals have a great capacity to convert resources into reproductive work, but they are limited by the rate of acquisition of resources for reproduction. They must spend most of their time foraging just to meet their high mass-specific maintenance needs. In contrast, large individuals have a great capacity to acquire resources, but they are constrained by the rate at which these can be converted into viable offspring.

The trade-off between these two limiting processes results in an optimal size. Since the bs in equations (5.10) and (5.11) are assumed to be the same for all organisms, the value of the optimum depends on taxon-specific constraints on structure and function that are expressed in the Cs. The values of C_6 and C_7 can be estimated from allometric equations for maximum productivity and maximum energy turnover, respectively.

We estimated the optimal body size for mammals, using data in Peters (1983) that suggest that $C6$ and $C7$ are approximately 6.0 W and 0.2 W respectively when mass is measured in kilograms. Substituting into equation (5.12) gives the distribution of reproductive power shown in figure 5.5. Note that this closely matches the size frequency distribution for mammals, and that it predicts an optimum body mass of approximately 100 g.

Two kinds of data from insular mammal faunas suggest that the optimal size for mammals is indeed about 100 g: (1) populations of species that on continents are larger than this size tend to evolve dwarf insular races, whereas populations of smaller species tend to evolve giant insular races (fig. 5.6a; Lomolino 1985); (2) as the area of a landmass and the number of species present decrease, the range of sizes represented in the fauna also decreases (Maurer, Brown, and Rusler 1992), so that on tiny islands where there is only a single species present, it tends to be close to the optimal size (fig. 5.6b; P. A. Marquet and M. L. Taper, unpub.). Thus, our model seems to be supported by data on the distribution of body sizes among mammal species on islands and continents. By defining fitness as reproductive power and predicting an intermediate optimal size, the model also makes testable predictions about variation in home range size and life history characteristics, which will be considered in chapter 7.

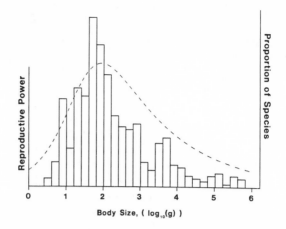

FIGURE 5.5. The distribution of reproductive power for mammals (dashed line) predicted from the model of Brown, Marquet, and Taper compared with the frequency distribution of body sizes among species (histogram bars, from fig. 5.2a). Note that the model predicts an optimal size of approximately 100 g. (From Brown, Marquet, and Taper 1993.)

One thing that the model does not do is to incorporate any interspecific interactions or any other environmental relationships that may not be mediated ultimately by energetics. Thus, while the model predicts an optimal size and fitness function (fig. 5.5), it does not by itself account for the assembly of a biota with the distinctive right-skewed distribution of sizes among the component species. It is easy to visualize how such assembly might occur. If there is only one species present, it should tend to evolve to the optimal size. Once this happens, however, the fitness function plotted in figure 5.5 will be altered, depressed at the optimum and exhibiting two new peaks on either side that should be exploited by the next two species to invade the community. It remains, however, to make explicit models of this process of faunal buildup, and to ask whether the assumptions and predictions of such models are supported by data.

ABUNDANCE

The Distribution of Abundance among Species

Frequency distributions of abundances among species, like the distributions of body sizes, have long intrigued quantitative ecologists. C. B. Williams (1964) collected and summarized enormous amounts of data, using light traps and other sampling methods, on the abundances of insect species at the Rothamsted experiment station and other locations in Britain. He got no less of a statistical expert than Sir Ronald Fisher to assist him

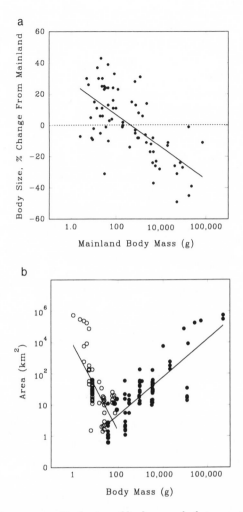

FIGURE 5.6. Two patterns of body sizes of land mammals that suggest that there is an optimal size of approximately 100 g. *(a)* Lomolino's comparison of the sizes of insular races (dots) with the sizes of their nearest mainland relatives, and the fitted regression equation (line) showing the microevolutionary trend toward increased size in small species and decreased size in large ones. (From Lomolino 1985.) *(b)* Pablo Marquet's plot of the sizes of the largest (solid symbols) and smallest species (open symbols) in the fauna of a continent or island as a function of the area of the landmass. Note that as the area—and thus the number of species in the fauna—decreases, the sizes of the remaining species tend to converge toward 100 g. (Adapted from Brown, Marquet, and Taper 1993.)

with analysis and interpretation of these data (Fisher, Corbet, and Williams 1943). At about the same time, Frank Preston (1948, 1962a,b), an American industrialist with a strong interest in birds and ecology, published a series of influential papers. MacArthur (1957) also published one of his first, most influential, and most controversial papers on this subject.

Williams, Preston, and MacArthur—and later Whittaker (1970)—were trying to characterize mathematically and understand another of the really general properties of ecological communities: they contain many relatively rare species and only a few very abundant ones. The four investigators used somewhat different mathematics to characterize the frequency distributions of abundance among species. Williams fitted his data with log-series distributions. In contrast, Preston argued that the data were better fit by a lognormal distribution with a specific relationship between the mean and variance that he called "canonical." MacArthur (1957, 1960b) advocated yet a third distribution function, the broken stick. Whittaker (1970) evaluated these different distributions, then followed a Japanese ecologist, Motomura (1932), in fitting some data to a geometric series. May (1975; see also Pielou 1977) presents a thorough treatment of the mathematical properties and possible ecological interpretations of all of these distributions.

There has been much debate (e.g., MacArthur 1966; Pielou 1966, 1977; Vandermeer and MacArthur 1966; May 1975; Sugihara 1980, 1989; Harvey and Godfray 1987; M. L. Taper and P. A. Marquet, unpub.) about which, if any, of these relationships best characterizes some general law about the distribution of individuals among species. Several authors have considered how resources might be apportioned among individuals and species (e.g., MacArthur 1957, 1960b; Sugihara 1980, 1989; Harvey and Godfray 1987; M. L. Taper and P. A. Marquet, unpub.). I will not go into these arguments here for four reasons. First, despite differences in the mathematics, the three distributions have generally similar form. Second, there seems to be little basis for distinguishing among the alternative distributions empirically. All of them provide reasonably good fits to many empirical data sets. At the level at which they might be differentiated, the data are often subject to artifacts of sampling. Choice of the spatial and temporal scale of the study, the organisms to be included, and the methods to be used in censusing or otherwise estimating abundance can all have effects that might bias the results toward one or another of the alternatives. Third, it seems unrealistic to try to explain the univariate frequency distribution of abundance and to ignore the effects of other variables. This may have been justified thirty years ago, but it is harder to defend in the light of recent advances. In addition to body size, it is important to consider the effects of variables, such as trophic status, that may or may not be held fairly constant by limiting the taxonomic scope of the study. Finally, and perhaps most importantly, a simple null model might explain these distributions. Since each species is rare in most of the places where it occurs and orders of magnitude more abundant in a few places (see chap. 4 and fig. 4.5), the distribution of abundances among species in a local community might simply represent a random sample of the abundances of the individual species.

For the moment, at least, it is sufficient to take the qualitative message of Williams, Preston, and MacArthur: virtually all ecological communities contain a few relatively common species and many rare ones.

Abundance and Body Size

One of the most obvious variables that affects abundance is body size. As outlined above, enough is known about the physiological consequences of body size to appreciate that it places severe constraints on the relative population densities of otherwise similar organisms living together in the same environment. Not enough is yet known, however, to interpret the variation that is observed empirically.

Several investigators have compiled data on abundance as a function of body mass and fitted them with standard allometric equations (e.g., Damuth 1981, 1987; Peters 1983; Peters and Raelson 1984; Marquet, Navarrete, and Castilla 1990). As mentioned above, however, these studies may be subject to sampling bias, and even if they are not, the fitted regressions usually account for only a small proportion of the wide variation in abundance that is typically observed. An alternative approach is to plot the data and try to explain the pattern of variation observed. Compilations of data for several different kinds of organisms suggest the general pattern depicted in figure 5.7 (Brown and Maurer 1987; Morse, Stock, and Lawton 1988; Lawton 1990). Several features of this relationship should be noted:

1. The highest population density is reached by species in relatively small, but not the smallest, body size categories. In fact, there appears to be a close correspondence between the size that attains the maximum density and that which contains the maximum number of species (see fig. 5.2).

2. From this highest-density size class, the maximum density in each size class tends to decrease regularly with increasing body size. The slope of a line fitted to these extreme values corresponds to an allometric exponent of -0.75 to -1.0.

3. From the highest-density size class, the maximum density in each size class also seems to decrease regularly with decreasing size. The slope of a line fitted to the extreme values is steep, with the corresponding exponent between 1.5 and 2.0.

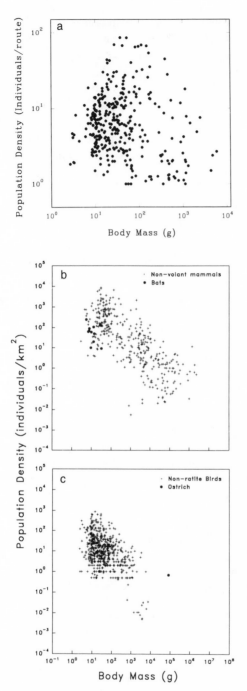

FIGURE 5.7. Distribution of population density as a function of body mass (both on logarithmic axes) for (a) 380 species of North American land birds, compiled from data from the North American Breeding Bird Survey by Brown and Maurer (1987); (b) 1,099 species of terrestrial mammals from around the world, from direct estimates of population densities compiled from the literature by M. Silva and J. H. Brown (unpub.); and (c) 999 species of land birds from around the world, from direct estimates of population densities compiled from the literature by M. Silva and J. H. Brown (unpub.). Note the similar shapes of the latter two distributions.

4. The minimum density is either independent of size (fig. 5.7a) or decreases with increasing size (fig. 5.7b,c).

Thus, rather than being tightly clustered around a line that would be well characterized by an allometric regression equation, the values tend to fall within a space that is approximated by a tetrahedron.

Brian Maurer and I (Brown and Maurer 1987, 1989) have suggested that many relationships between ecological variables for multiple species are of this general type. There is too much variation to characterize the distribution with a line or curve, but the variation appears to be restricted to a well-delineated space. Often, if the axes are scaled appropriately, the space can be characterized as a polyhedron by drawing lines separating regions of parameter space containing points from regions that are empty. We call these constraint lines. They imply that it is at least theoretically possible for a species to exhibit any combination of values that fall within the space, but impossible (or at least highly unlikely) that it can take on values that would lie outside the space.

This framework is potentially useful, but it needs to be developed further. Other investigators have begun to develop rigorous statistical procedures to define the constraint lines that best fit empirical data (e.g., Blackburn, Lawton, and Perry 1992). Delineating the constraint space places the focus on explaining the pattern of variation rather than on accounting for some average trend in the data. Since each of the constraint lines potentially reflects the influence of a different limiting process, the investigator is challenged to develop and test a hypothesis for each one.

In the case of population density and body size (fig. 5.7), the allometric relationships discussed above offer hypotheses for the upper limit on abundance as body size both increases and decreases from the size class of maximum abundance. Some other structural/functional constraint apparently determines the minimum size of the group; this has been the subject of much discussion in comparative physiology (see Calder 1984; Schmidt-Nielsen 1984).

It remains to explain the lower limit on abundance as a function of body size. Most biologists would probably predict that minimum density decreases with increasing body mass (a constraint line with negative slope), for several reasons. First, since mobility usually decreases with decreasing size, small organisms cannot become as rare as their larger relatives and still find mates. Second, if minimum viable population size, the total number of individuals required to have some reasonably low probability of extinction, is independent of body size, larger organisms will tend to have lower densities by virtue of their larger home ranges. Further, if small organisms tend to have higher variances in population growth rates, they may have higher probabilities of extinction for a given population size and thus higher minimum viable population sizes. Data on mammals (fig. 5.7c; see also Damuth 1981, 1987, 1991; M. Silva and J. Downing, pers. comm.) appear to show that the density of the rarest species in each size category does indeed decrease with increasing body size.

On the other hand, some empirical data for birds suggest that minimum density may be nearly independent of body size (fig. 5.7a; see also Lawton 1990; Nee et al. 1991). I suggest that there are two reasons for this. First, birds are exceptional for their mobility, and the capacity to disperse long distances may enable even the smallest species to be as sparsely distributed as their larger relatives and still find mates and avoid extinction. The densities of birds are, on average, much lower than those of mammals of comparable size (see chap. 7). Second, although some small birds may appear to have crude densities (i.e., the same number of pairs per unit area, especially when measured at a large scale) as low as those of much larger birds, they probably have higher "ecological densities" (sensu Damuth 1987). That is, the smaller species tend to have smaller home ranges and to use the environment in a much more patchy way.

Both of these explanations suggest that the apparently low densities of some of the smallest birds are a somewhat artificial consequence of the scale of sampling. This is supported by a recent study by Marina Silva and myself (M. Silva and J. H. Brown, unpub.), in which we collected data on the ecological densities of hundreds of bird and mammal species, taking care to insure that the values reflected the scale at which foraging individuals used the environment. Our results showed that minimal density in both birds and mammals appears to scale approximately as $M^{-0.75}$ (fig. 5.7b,c). I suspect that this is the general pattern.

Before leaving this topic, it seems reasonable to ask what the pattern of abundance as a function of body mass would look like if all organisms, from microbes to the largest plants and vertebrates, were considered. Sufficient data for a definitive answer do not exist, so we are free to speculate. I predict that the relationship would resemble qualitatively the pattern depicted in figure 5.8. Even if I am correct, however, several questions remain. What size category and kind of organism attains the highest population density? Good candidates would seem to be some species of insect, nematode, or microbe. What size category exhibits the greatest variation in abundance? How rapidly does the minimum density decrease as body size increases? Probably the rarest small organisms are highly specialized pathogens and parasites, which, like birds, have attributes that enable them to find mates and suitable environments (hosts) at very low density.

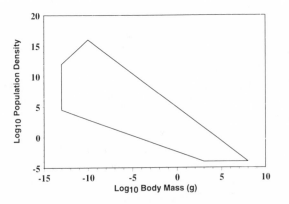

FIGURE 5.8. The hypothetical constraint space enclosing the distribution of local abundance as a function of body mass in all organisms. Note that population density is hypothesized to vary over many orders of magnitude in very small organisms.

How do the abundances of such parasites compare with those of their orders of magnitude larger hosts? Unless they are proportionally more abundant, minimum population density may decrease only moderately with decreasing body size, and the greatest variation in abundances may occur in species of very small size.

ENERGY USE: EFFECTS OF BODY SIZE AND ABUNDANCE

Energy Use of Species

So far, we have been concerned with the distributions of body sizes and abundances among species. Now let's consider the ecological effects of species of different sizes. What kinds of organisms of what body sizes use most of the energy and nutrients available within a habitat? This question is of considerable ecological interest because it addresses the roles of different species in biogeochemical ecosystem processes. Conventional ecological wisdom would suggest that species of small body size have the greatest impact.

This is an empirical question, but the theoretical framework developed above can help to place it in conceptual perspective. The rates of energy and nutrient use per unit area for each species within a local community (L) will be proportional to $E \cdot N$, the average metabolic rate (E) per individual times the population density (N). The metabolic rates of related species are closely and positively correlated with individual body mass (eq. [5.1]; Peters 1983; Calder 1984; Schmidt-Nielsen 1984). As I have just shown, however, there is a great deal of variation around the generally negative relationship between population density and body size. Thus, energy use per unit area of species populations cannot be characterized accurately with a simple allometric equation.

It is possible, however, to say something about the maximum rates of energy and nutrient use as a function of body size. These will be determined by the population densities of the most abundant species in each size class, which vary with body mass according to equation (5.5) or (5.6). Therefore the upper limit on $E \cdot N$ should vary between $M^{0.0}$ and $M^{0.25}$. Damuth (1981, 1987; see also Van Valen 1973a, 1976; Peters 1983) has emphasized the former value, at least for animals. He has gone so far as to suggest that the energy use of all species, not just the most abundant species in each size class, within an assemblage tends to be equal. If this were true, it would be an elegantly simple law of nature.

Unfortunately, however, both data and theory suggest that it cannot be correct. For energy use of all species to be equal requires that abundances scale closely as $M^{-0.75}$. Even Damuth's (1987) own compilations of data for mammals show that there are two to three orders of magnitude variation in population density (and hence in energy use per unit area) among species of similar body size. The range is at least an order of magnitude greater if data for both birds and mammals are plotted on the same graph (fig. 7.6). Further, it is clear that the rare parasites and pathogens discussed above must use only a tiny fraction of the energy required to sustain their hosts.

Brian Maurer and I (Brown and Maurer 1986; see also Nee et al. 1991) assembled data on the densities of several kinds of organisms within local communities and then used the allometric relationship between metabolic rate and body mass within the taxon (eq. [5.1])[4] to estimate rates of energy use per species. These studies indicate that there is a great deal of variation in energy use, much of it among species of similar body size. Nevertheless, there is a consistent tendency for the larger species to have higher values. This can probably be attributed to two things. First, all of the largest species tend to have relatively high rates of energy use. As suggested by figure 5.7, large organisms are constrained to a narrow range

---

4. Direct measurements of metabolic rates—especially of ecologically relevant metabolic rates under field conditions—are not available for many species. Because of the high correlations between log E and log M within a taxon, however, metabolic rate can be estimated relatively accurately using equation (5.2). In the original study, Brown and Maurer (1989) assumed that existence energy requirements in the field varied as $M^{0.67}$, rather than as $M^{0.75}$; the latter is probably more accurate. This does not alter the conclusions, however, because using a larger value for the exponent only increases the tendency of larger species to have higher values of $E \cdot M$. Even for morphological and physiological variables, however, there is considerable variation among species around the fitted allometric equations. A better method would be to obtain good data for each species on such variables as field metabolic rate, home range size, and population density (P. A. Marquet et al., unpub.).

of low densities, and these values, when multiplied by the high metabolic rates per individual, give relatively high values for all large species. Second, the smallest species often have relatively low population densities, much lower than predicted by equation (5.6) (fig. 5.7). These low values, when multiplied by the low metabolic rates per individual, give low estimated rates of energy use for many of the smallest species. We conclude that, on a per species basis, organisms of large body size on average have higher rates of energy and nutrient use than their smaller relatives.

This does not mean, however, that no small species use as much energy as their larger relatives. In fact, because of the wide variation in population densities among small organisms, the most abundant species in each size class (at least down to approximately the modal size class in fig. 5.7) tend to use energy at approximately equal rates (fig. 5.9). If maximum population density (N) scales as $M^{-0.75}$ and metabolic rate scales as $M^{0.75}$, then $E \cdot N$ for the most abundant species in each size class must scale as $M^{0.0}$.

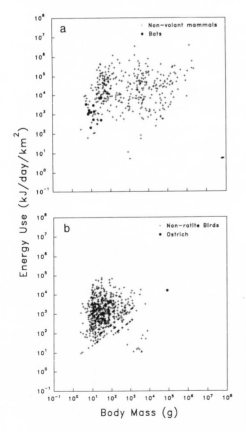

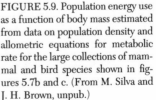

FIGURE 5.9. Population energy use as a function of body mass estimated from data on population density and allometric equations for metabolic rate for the large collections of mammal and bird species shown in figures 5.7b and c. (From M. Silva and J. H. Brown, unpub.)

Energy Use of All Species in a Size Class

In the beginning of this chapter, I showed that in most large taxonomic assemblages there are more species of small than of large body size. Do rates of energy use by all of the small species taken together equal or exceed those of their larger relatives? This question can be addressed by summing the values of $E \cdot M$ for all of the species within a logarithmic size class. When this is done for the data on North American land birds (using the population density values from the BBS graphed in figure 5.7a), rates of energy and nutrient use appear to be approximately equal for size categories from the mode to the largest (fig. 5.10), and these are all substantially greater than the values estimated for the smallest size classes (Maurer and Brown 1988).

The conclusion drawn from these analyses, then, goes counter to conventional ecological wisdom. The flow of energy and materials through

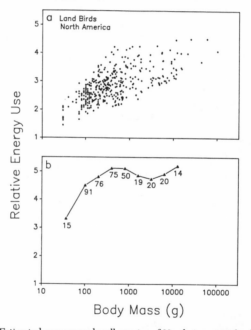

FIGURE 5.10. Estimated energy use by all species of North American land birds, plotted separately (*a*) and then summed for all species within equal logarithmic size classes (*b*, connected line, with the number of species in each category indicated.) Note that as size increases, the summed values for size categories increase up to about 100 g and then remain nearly constant. These values were calculated by Maurer and Brown based on data from the Breeding Bird Survey; if population densities of the least abundant small species are systematically underestimated by the BBS (see text and fig. 5.9), this should increase the values for individual species in the lower left part of the graph but not much affect the total energy use for these size categories. (From Maurer and Brown 1988, fig. no. 3 on p. no. 1928.)

ecosystems is not dominated by small organisms. If anything, the smallest organisms use energy and nutrients at lower rates than their larger relatives. Over a wide range of sizes, however, energy use and nutrient exchange is largely independent of body size, as Damuth (1981, 1987) suggested. Note, however, that all of the analyses have been confined to organisms within a restricted taxonomic group, such as a single class of vertebrates. What if all organisms, or at least, say, all animals, vertebrate and invertebrate, endotherms and ectotherms, were considered together? Would the energy flow through terrestrial ecosystems be dominated by the large vertebrates or by the small insects? Would respiration in pelagic marine ecosystems be greater through fishes, mammals, and birds than through planktonic crustaceans? I don't think that we have sufficient data to answer these questions.

If I make a prediction, however, perhaps someone will be challenged to test it. I suspect that even in these cases, provided reasonably accurate measures of energy use are obtained and summed over all organisms (or at least all animals) in logarithmic body size categories, the values for the smaller size classes will not more than equal those for the larger ones. One often hears that microbial decomposers or herbivorous insects are the dominant heterotrophic organisms in terrestrial ecosystems, but I suspect that the vertebrates[5] are just as important in the transfer of energy and materials.

In all of the foregoing discussion I have been concerned with the rates of energy use per unit area (L) of different-sized species within local communities. What about the total energy use (G) of different species in the entire world? Such energy use will be approximately $E \cdot N \cdot A$, where E is metabolic rate per individual, N is average local population density, and A is the area of the geographic range. Since the size of the geographic range of large animals tends to be at least as large as that of small ones (see fig. 6.3), total energy use per species will tend to increase even more rapidly with increasing body size than does energy use within local communities. The dominant animal consumers on the earth today are *Homo sapiens* and the large domestic animals (see chap. 12).

A Caveat

Before leaving this topic, one important point of qualification should be made. Care must be exercised in inferring rates of energy and nutrient transfer in ecosystems from metabolic rates of individual organisms. For

5. In most terrestrial ecosystems the dominant vertebrate consumers are herbivorous mammals. In most ecosystems the native megaherbivores have been reduced or extirpated by humans, but they have to some extent been replaced by domestic livestock (see chaps. 11 and 12).

example, in terrestrial ecosystems endothermic birds and mammals have metabolic rates one to two orders of magnitude greater than ectothermic reptiles of similar body size. The vast majority of food consumed by birds and mammals is metabolized to support maintenance of their high body temperatures and levels of activity; only a small fraction is allocated to growth and reproduction. In contrast, reptiles use a much smaller fraction of the energy and nutrients they consume for maintenance and a correspondingly larger fraction for production. As a consequence of their high food requirements, endothermic birds and mammals are potentially important as consumers in ecosystems; when abundant, they have a major impact on lower trophic levels, but produce relatively little biomass for their own predators to consume. Reptiles, by contrast, are potentially much more important as producers; for the same amount of food consumed, they can attain much higher population densities (by at least one order of magnitude) and thus provide a much larger food resource for organisms at higher trophic levels than birds or mammals do (Turner 1970; Pough 1980, 1983).

This example makes one other point. There is a great deal of variation among organisms that is not only interesting but also has important ecological, biogeographic, and evolutionary implications. The magnitude of this variation is limited, and many potentially important variables are in a sense "controlled," when comparisons are restricted to sufficiently closely related species within taxonomic groups. This variation needs to be taken into account, however, when comparisons are extended to include distantly related kinds of organisms.

CONCLUDING REMARKS

In this chapter I have considered variation among species in body size, abundance, and energy use. There are tantalizing hints that the statistical patterns I have documented in birds and mammals might be very general, occurring in all or nearly all organisms. There is no a priori reason to believe that the patterns and processes should be specific to organisms that live on land, weigh more than 2 g, are endothermic, and have backbones. On the other hand, most of my data and insights come from birds and mammals. Do aquatic animals, land plants, and microbes exhibit similar patterns? I would encourage others to collect the data and do the analyses. Any phenomenon in biology that is apparently so universal should elicit great interest.

Body size and energetics provide a promising vehicle for making connections between macroecological patterns and underlying mechanisms at the level of individual organisms. Body size places severe constraints on morphology, physiology, and behavior. Allometric equations describe these general empirical relationships at the level of the individual. They provide a basis for making mathematical models of the ecological effects of body size on use of food resources and space by individuals, on the density, dispersion, and demography of populations, and on the transfer of energy and nutrients within ecosystems. These size-specific attributes of species, in turn, have important biogeographic, macroevolutionary, and practical, conservation-related implications that will be considered in later chapters.

6 The Assembly of Continental Biotas

Geographic Range

Now, having analyzed the attributes of individual organisms that affect their assembly into biotas on different spatial scales, I will consider how species are distributed across the landscape. In particular, I will focus on the variation among species in the sizes, shapes, and locations of their geographic ranges.

SIZE OF THE GEOGRAPHIC RANGE

Distribution of Range Size among Species

Well before Hutchinson and MacArthur called attention to the distribution of body sizes and Williams, Preston, and MacArthur analyzed the distribution of abundances, a biogeographer named John Willis (1922) plotted frequency distributions among species of the areas of geographic ranges and noted that different taxonomic groups exhibited a common pattern (see fig. 2.1). Willis called these distributions "hollow curves" because of their shapes when plotted on linear axes. A somewhat different pattern is obtained if the data are plotted on logarithmic axes (fig. 6.1), but there are clearly many species with small ranges and only a small minority with very large ones. Does this sound familiar?

As implied by the title of his book, *Age and Area,* Willis hypothesized that these distributions reflected the history of the species. He thought that species tended to originate with small ranges and expand their distributions, so that the older species inhabited larger regions. Although we would now dismiss such a suggestion of an orthogenetic evolutionary trend as a quaint anachronism, we cannot dismiss the pattern that Willis documented. Analyses of contemporary data give the same result (fig. 6.1; Anderson 1977; Rapoport 1982; Pagel, Harvey, and Godfray 1991). This

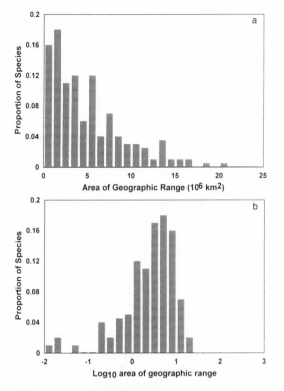

FIGURE 6.1. Frequency distribution of number of species as a function of areas of their geographic ranges for North American land birds plotted on (a) arithmetic and (b) logarithmic axes. Note that the arithmetic plot shows the same "hollow curve" shape described by Willis (1922; see fig. 2.1). The majority of species have small ranges that represent only a small fraction of the continental area.

frequency distribution appears to be another of those very general emergent statistical characteristics of taxonomic assemblages. Like similar patterns of body size and abundance, it appears to reflect yet another fundamental way that species within taxonomic groups have divided up the earth's resources, in this case by inhabiting different amounts of geographic space.

Consider the implications. For one thing, the pattern implies that related species differ widely in their ecological requirements. Species with small ranges must be limited by their narrow tolerances for some abiotic or biotic conditions, while those with very large ranges must be able to tolerate a wide range of abiotic conditions, to coexist with many other species, and probably to use a wide range of resources. This hearkens back to themes discussed in chapter 4, and I shall return to some of them shortly. The pattern also implies that continental biotas are assembled from these different kinds of species in a complex, but at least somewhat predictable, way. How the many small ranges and relatively few large ones are distributed in geographic space and how they overlap with one another determines in large part the variation in the diversity and other characteristics of biotas across the continental landscape.

Range Size and Abundance

From what has been said in the previous two chapters, I would expect that variation in the sizes of geographic ranges is correlated with, or at least constrained by, both abundance and body size. This appears to be the case, although unfortunately I know of only one example: analysis of a large data set on North American land birds (Brown and Maurer 1987). A somewhat similar analysis of Australian land birds is probably not comparable because the estimates of abundance have been standardized only within, not across, species (Schoener 1987, 1990).

In chapter 4 I have shown that among closely related and/or ecologically similar species there tends to be a positive correlation between local population density and extent of spatial distribution at the scale of geographic range. Species with large ranges tend to be more abundant throughout those ranges than species that are restricted to small areas. When many species in a large taxonomic group, such as North American land birds (fig. 6.2), are considered, however, the pattern of variation is more complex than would be indicated by fitting a linear regression (to the logarithmically transformed data). While there is a great deal of variation, it appears to be constrained. Two constraint lines are quite apparent. One is the limit on the maximum size of the geographic range, which simply reflects the area of the North American continent. (Some species actually have larger ranges than are shown because their distributions extend into South America or Eurasia). A second constraint line reflects the apparently constant minimum population density shown in figure 5.7a. In addition to these two obvious constraints, there is a very low density of points in the upper left-hand portion of the graph: only a very few species (2 or 3 out of 380) are very abundant locally but have small geographic ranges.

Brian Maurer and I have suggested that this third limit on the variation represents a relative, rather than an absolute, constraint. Absolute constraints represent impossible combinations, whereas relative constraints represent improbable combinations. In the present case, is possible for a species to have niche requirements that enable it to be very abundant locally, yet restricted to a small geographic range. As pointed out in chapter 4, the bird species that have these attributes are confined to a habitat that is highly productive of suitable food. A prime example is the tricol-

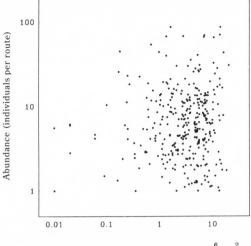

FIGURE 6.2. Average local abundance as a function of area of geographic range for North American land birds. Note that the data points tend to fill in a roughly triangular space in the lower right part of the graph, although there are some conspicuous outliers (see text). (After Brown and Maurer 1987, based on data from the Breeding Bird Survey.)

ored blackbird (*Agelaius tricolor*), which occurs in the marshes of the Central Valley of California and some nearby areas.

Despite these constraints, there is much more variation when all North American birds are combined together than when only closely related or ecologically similar species are considered. This is not surprising, because each taxonomic or ecological subgroup is subject to additional constraints. For example, both small songbirds (order Passeriformes) and much larger diurnal birds of prey (order Falconiformes) show reasonably high correlations between local abundance and area of geographic range when considered separately, but the densities of small passerines are much higher on average than those of large, carnivorous hawks. Thus, the pattern for a large taxonomic assemblage, such as the North American land birds shown in figure 6.2, can be visualized as being made up of many smaller patterns representing different taxonomic groups (or ecological guilds), each of which exhibits a different, more or less linear relationship with substantial variation around it (fig. 6.3).

Range Size and Body Size

Area of geographic range should also be constrained by body size, in part because body size constrains abundance, which in turn constrains geo-

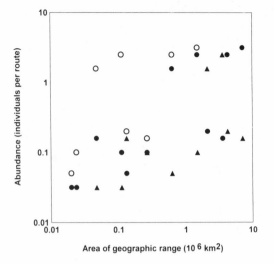

FIGURE 6.3. Average local abundance as a function of area of geographic range (both on logarithmic axes) for three taxonomic and ecological groups of North American land birds: open circles, wood warblers (subfamily Parulinae); solid circles, sparrows (subfamily Emberizinae); triangles, diurnal birds of prey (order Falconiformes). Note that there is a positive correlation between abundance and area of range for each group, but that the warblers tend to have the highest and the hawks the lowest abundances. (Data from the Breeding Bird Survey.)

graphic distribution, as I have just shown. Plots of area of geographic range as a function of body mass on logarithmic axes provide evidence of such constraints. In this case I have analyzed data for both birds and mammals in North America, and the two taxa exhibit similar patterns of variation (fig. 6.4). At least three and perhaps four constraint lines can be fitted to these data. Two of these, the structural/functional limits on minimum size of birds and mammals and the trivial upper limit on the maximum area of geographic range placed by the size of the continent, are obvious, absolute constraints.

There is also a limit on the minimum area of the range, and this is of considerably more interest. For one thing, it is only a relative, not an absolute, constraint. For another, it is not clear whether there is just one constraint or two. In the data for both birds and mammals there are few species of large body size with small geographic ranges. Presumably, this reflects the effect of body size on population size and thereby on probability of extinction. Species of large size are constrained to have low population densities (see above), and as a result total population size will be small unless the individuals are distributed over large areas. Probability of extinction is greatly increased at small population sizes (e.g., MacAr-

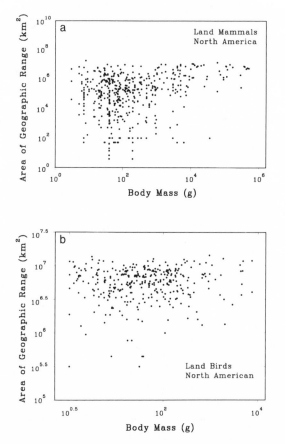

FIGURE 6.4. Area of geographic range as a function of body mass (both on logarithmic axes) for the species of North American terrestrial mammals *(a)* and land birds *(b)*. Note that both kinds of organisms exhibit similar patterns, with the species of large size (and perhaps also of very small size) tending not to have small ranges.

thur and Wilson 1967; Goel and Richter-Dyn 1974; Goodman 1986) and it may be additionally increased by restriction to a small area where the entire species could be eliminated by a regional abiotic or biotic catastrophe (e.g., a volcanic eruption or an outbreak of disease). The constraint set by extinction is relative, however. It is not impossible for a species of large size to have a small geographic range. It is just unlikely that it will be able to persist for long in that condition; unless it is able to increase its numbers and expand its distribution, it has a high probability of going extinct. Indeed, many of the species with small ranges for their sizes are listed as rare and endangered. The California condor, one of the largest

North American land birds, was confined to a small region of southern California before the remaining individuals were captured for a zoo breeding program.

There is also a hint in the data for both birds and mammals (fig. 6.4) of an additional constraint. The minimum area of geographic range appears to be consistently smaller for the very smallest species than for their somewhat larger relatives. This is consistent with the suggestion (above) that the minimum size of the range is determined primarily by the probabilistic effect of low population size on the probability of extinction, and also by the fact that the smallest organisms are constrained to have low minimum population densities (see fig. 5.7b,c).

GEOGRAPHIC RANGES ON CONTINENTAL LANDSCAPES

The Shapes of Ranges

As a variable, area of the geographic range has many of the same advantages and disadvantages as population density. On the positive side, both are useful quantitative attributes of species that are easy to obtain from published sources. On the negative side, both provide only one simplistic measure of the complex spatial distributions of individual organisms. The area of the range is usually obtained by measuring with a planimeter (now with a digitizer interfaced with a computer) the area within published range maps and making the appropriate correction for the scale of the map. The one value thus obtained for each species does not include any information on the shape, location, internal structure, or dynamics of the range.

Some measures of shape and location of the range are relatively easy to obtain from published range maps, and they offer additional insights not available from measures of area alone. The data on areas raise interesting questions about how geographic space is allocated among species: where ranges of different sizes are located, how they overlap with one another, and how they are related to the variation in physical conditions, such as features of geology and climate. These questions can be addressed, at least in a preliminary way, by some very simple measurements and analyses. Unfortunately, I have done only a little preliminary work, but the results raise interesting questions and suggest additional analyses.

Suppose that we want to quantify the shape, as well as the size, of the range. There are a number of ways to do this. Rapoport (1982) analyzed the ratio of the perimeter to the area. He found that among North American land mammals the perimeter/area ratio averaged about 10 and did

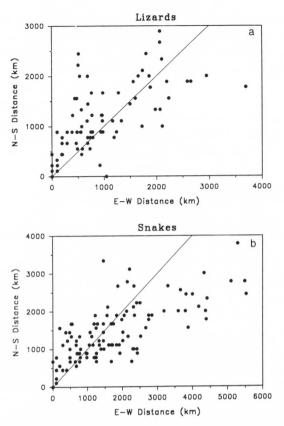

FIGURE 6.5. A plot of the north-south versus the east-west dimensions (in km) of the ranges of North American lizard *(a)* and snake *(b)* species. The line represents the configuration of hypothetical (e.g., circular) ranges in which the two dimensions are equal. Note that for both kinds of reptiles, the small ranges tend to be oriented north-south (above the line), whereas the large ranges tend to be longer in the east-west dimension (and hence below the line). (Data from R. Vestal, unpub.)

not seem to vary with the size of the range. This finding suggests that ranges might have some interesting fractal properties.[1]

I have done something even simpler. I measured the greatest north-south and east-west dimensions of the range. These values can then be plotted, one on each axis, to provide a first approximation of both size and configuration of the range (fig. 6.5). Note that in such a graph, ranges that are square, circular, or any other shape in which the north-south and east-

1. One needs to be careful, however, because the fractal properties may be in the minds of the cartographers who drew the range maps, rather than in the distributions of the organisms themselves.

west dimensions are equal will fall along a line of equality, with small ranges near the origin and larger ones farther away. Ranges that are elongate in a north-south direction will fall above the line of equality, and those that are attenuated east-west will plot below the line.

The first such graphs were produced for North American reptiles as part of a class project by Renée Vestal, a student in my biogeography course. Lizards and snakes, graphed separately, show the same qualitative pattern (fig. 6.5a,b). The small ranges lie predominantly above the line; they tend to be attenuated in a north-south direction. In contrast, the large ranges tend to be below the line, so they have significantly greater east-west than north-south dimensions.

It may seem surprising—it still does to me—that such a simple analysis can reveal such a clear and repeatable pattern. Once the pattern is revealed, however, it is not difficult to come up with hypotheses to explain it. Species with small geographic ranges must be limited by environmental conditions that vary on a local or regional scale. A glance at a map of the physical geography of North America immediately shows that the major mountain ranges, river valleys, and coastlines tend to be oriented predominantly north-south. Associated with these geological features is regional variation in soils, climate, and vegetation that can be hypothesized to determine the boundaries of small ranges. Species with large ranges, on the other hand, are distributed over much of the continent, so their ranges cannot be so affected by such regional geographic features. I hypothesize that they are limited primarily by large-scale patterns of climate and vegetation, and these tend to run east-west in broad latitudinal bands, like the colored climatic zones that are printed on seed packets and in gardening books. Perhaps it is not surprising that widely distributed ectothermic reptiles are sensitive to such climatic variation.

Here is a case in which it is possible to test macroecological hypotheses without performing manipulative experiments. First of all, if this pattern really does reflect the effect of the geography of North America on the configurations of geographic ranges, then other kinds of organisms should show similar patterns. Similar analyses can easily be done using data from published range maps of North American birds and mammals. If my hypotheses are correct, the small ranges should show the same north-south pattern, but will endothermic birds and mammals show the same east-west pattern as presumably more temperature-sensitive ectothermic reptiles? The data (fig. 6.6a,b) indicate that they do. In fact, it is surprising how similar the patterns for the three groups of terrestrial vertebrates are. Among other things, this implies that the distributions of birds and mammals, while they might or might not be so directly influenced by en-

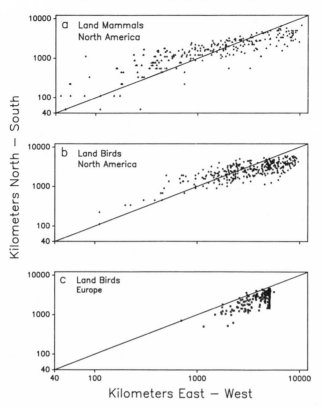

FIGURE 6.6. Plots, similar to those in figure 6.5 but on logarithmically scaled axes, showing the north-south and east-west dimensions of the ranges of (a) North American terrestrial mammals, (b) North American land birds, and (c) European and North African land birds. Note that all faunas show similar patterns, except that the small as well as the large ranges of the European birds tend to be oriented east-west. These patterns are hypothesized to reflect the main geographic and climatic features of the two continents. (Figure reprinted from J. H. Brown and B. A. Maurer, "Macroecology: The Division of Food and Space among Species on Continents," *Science* [1989] 243: 1145–50. (c) 1989 AAAS.)

vironmental temperature, are ultimately just as affected by large-scale climatic variation as the ranges of ectothermic reptiles.

I have gone a bit further. If the hypothesis that the north-south orientation of small ranges is caused by the regional-scale geographic features of North America is correct, then small ranges should have a different orientation on a continent with different geography. In Europe and North Africa, the Mediterranean Sea, major mountain ranges (Alps, Pyrenees, Atlas), and associated regional habitats are aligned predominantly east-west, so my hypothesis would predict that small geographic ranges should be oriented similarly. Figure 6.6c shows that this prediction is upheld for European birds; virtually all points are below the line of equality.

Another interesting feature of the ranges of European birds is that the smallest of them are considerably larger than the smallest ranges of North American species (compare fig. 6.6b and c). This difference has at least two possible explanations. According to one hypothesis, the east-west orientation of geographic features in Europe may have contributed to the extinction of species with small ranges during the Pleistocene. As the climate fluctuated and glaciers advanced and retreated, species would have tended to respond with latitudinal shifts in their ranges, but those with small ranges would have been pushed up against the mountains and other barriers, increasing their probability of extinction. This would have been much less likely in North America, where the barriers are aligned north-south and the species would have been relatively free to shift their ranges latitudinally. Such an effect of Pleistocene climate interacting with European geography has been proposed to account for the lower diversity of trees in Europe than in eastern North America (MacArthur 1972b). Another hypothesis, also historical, is that the long history of human habitation and conversion of forests and wetlands to agricultural fields in Europe has caused the extinction of species with small ranges. These hypotheses are hard to evaluate, not mutually exclusive, and probably not the only ones that might be proposed.

Sizes of Ranges and Latitude

Another pattern of distribution of range sizes concerns the latitudinal placement of ranges of different size. At least in North America, there is a general tendency for range size to decrease with decreasing latitude (fig. 6.7). This appears to be another case, however, in which the pattern of variation is much more informative than the average trend.

George Stevens has called attention to one aspect of this pattern that he calls "Rapoport's rule." As first Rapoport (1982) and then Stevens (1989; but see Rohde, Heap, and Heap 1993) noted, species whose ranges are centered at increasingly high latitudes tend to be distributed over an increasingly wide range of latitudes (fig. 6.8). Stevens (1992) has shown that a pattern exactly analogous to Rapoport's rule holds for elevational distributions: within the same latitude, the elevational ranges of species increase with the elevation of the midpoint of their ranges (fig. 6.9). This pattern is very general; it holds for organisms from land mammals to coniferous trees. It clearly begs for a mechanistic explanation. I can suggest five hypotheses.

The first hypothesis is that both Rapoport's rule and the general tendency of small ranges to be concentrated at low latitudes is just a conse-

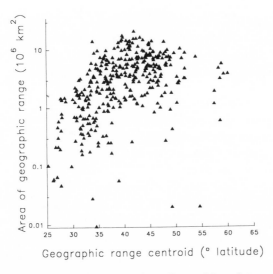

FIGURE 6.7. Area of geographic range (on a logarithmic scale) as a function of the latitude of the center of the range for North American land birds. Note that while there is considerable variation, there is also a pronounced tendency for the sizes of the ranges to decrease from the high latitudes toward the tropics. (Courtesy of B. Maurer.)

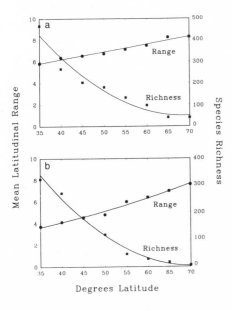

FIGURE 6.8. The relationship between latitudinal breadth of the geographic range and the median latitude of the range for (a) marine mollusk species along the coasts of North America and (b) tree species in the United States and Canada. Note that both taxa obey Rapoport's rule: the species at higher latitudes are distributed over a wider range of latitudes than their tropical relatives. Species richness is also plotted as a function of latitude, and exhibits the opposite pattern from latitudinal range. (Compiled from data supplied by G. Stevens.)

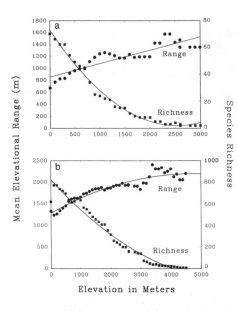

FIGURE 6.9. The relationship between altitudinal breadth of the geographic range and the median altitude of the range for *(a)* tree species in Costa Rica and *(b)* bird species in Venezuela. Species richness is also plotted. These are examples of Rapoport's rule applied to elevation instead of to latitude. Note that for both trees and birds, elevational range increases as species diversity decreases with increasing elevation. (After Stevens 1992.)

quence of the geometry of North America: because the continent tapers from north to south, the physical space available to accommodate species becomes increasingly limiting at low latitudes and therefore ranges are smaller. I think that this hypothesis can be rejected as a general explanation, because the trend is clear well before the latitudinal breadth of the continent begins to narrow conspicuously from northern Mexico to Panama. Furthermore, it cannot account for the elevational version of Rapoport's rule, because available area obviously decreases with increasing elevation.

Stevens suggests a second hypothesis: that spillover or dispersal between source and sink habitats can play a greater role at lower latitudes, where abiotic conditions tend to be more favorable and less variable in space and time. Consequently, at low latitudes species with small ranges and small populations are able to avoid extinction because individuals and populations are able to survive for long periods in sink habitats where birth rates are on average only slightly lower than death rates. Thus there would be a large reservoir of individuals in sink habitats available to recolonize the more favorable source habitats and reestablish populations after local extinctions (see Holt 1983; Pulliam 1988). Furthermore, subtle environmental changes might cause small changes in relative birth and death rates sufficient to convert former sink habitats into source habitats and vice versa.

The third hypothesis comes from an insight of Daniel Janzen. In a provocative paper entitled "Why Mountain Passes Are Higher in the Tropics," Janzen (1967) noted that comparable differences in elevation ought to present more severe barriers to dispersal at low than at high latitudes. The reason is simple: the seasonality of environmental temperature decreases with latitude. At high latitudes, a species confined to a low-elevation zone of climate and vegetation could disperse over mountain passes during the summer without encountering abiotic conditions any more severe than it experiences in the lowlands during the winter (and vice versa for a montane species crossing the lowlands in winter). In contrast, a tropical species crossing a barrier of similar elevation would have to tolerate conditions that it had never experienced, no matter what time of year it dispersed. Janzen and Stevens suggest that the greater severity of barriers at lower latitudes contributes to the formation and persistence of narrowly endemic species.

I can offer a fourth hypothesis that invokes extinction caused by the climatic fluctuations during the Pleistocene. We know that large areas of northern North America (and also Eurasia, but not the southern continents) were covered by vast ice sheets during the glacial periods. The boreal and north temperate organisms that survived the more than twenty glacial/interglacial cycles of climate during the last 2 million years could only have done so by repeatedly shifting their ranges hundreds or even thousands of kilometers, many degrees of latitude, and many meters of elevation to remain in environments that offered generally similar conditions. Many species may have gone extinct because they were unable to make these shifts. Such environmental changes would tend to differentially affect species with narrow requirements and restricted geographic ranges because they would experience frequent local extinctions and have to disperse long distances to avoid global extinction. As we learn more about the Pleistocene climate and vegetation in the tropics, it is becoming apparent that they also exhibited substantial changes. It is also apparent, however, that the biogeographic effect of these changes was much less: species could have survived by making modest horizontal and elevational shifts in their ranges. Thus I suggest that the climatic cycles of the Pleistocene have selected for broadly tolerant, wide-ranging species at high latitudes.

A final hypothesis is based on the observation that range size and species diversity exhibit opposite latitudinal trends: range size decreases while species diversity increases from the poles toward the equator (fig. 6.8; and similarly for elevation, fig. 6.9). This suggests the possibility that at lower latitudes the ranges of species tend to be increasingly limited by interactions with the increasing number of other species with which they come into contact. This intriguing idea, which I attribute to Dobzhansky (1950) and MacArthur (1972b), will be discussed in more detail in chapter 8 when I consider population processes and interspecific interactions.

Three comments should be made about these hypotheses. First, as I have often noted before, they are not mutually exclusive. Some combination of two or more of these phenomena may well be necessary to account for Rapoport's rule. Second, it may not be so difficult as it might at first appear to evaluate these hypotheses. For example, the effect of continental geometry can be assessed more rigorously by comparing the distributions of the same kinds of organisms in North America, which narrows from the Arctic to the tropics, and in South America, which tapers rapidly from the tropics to cold temperate latitudes. The hypothesis about the historical legacy of Pleistocene glaciation and climate can be evaluated by comparing patterns in northern North America and Eurasia, where continental ice sheets and climate changes were extensive, with those for the temperate zones of southern continents, where the magnitude and effects of Pleistocene cycles were probably more comparable to those in the tropics. Finally, the similarity between the latitudinal and elevational patterns suggests that they may be caused by similar mechanisms. I will say more about this in chapter 8.

Whatever mechanisms ultimately turn out to underlie Rapoport's rule, the fact that species with small geographic ranges apparently originate and persist more readily at low than at high latitudes must contribute importantly to the high diversity of species in the tropics. Any effort to provide a complete and general explanation for the almost universal latitudinal gradient in species diversity must include a macroecological perspective and consider how the sizes, configurations, and overlaps of geographic ranges vary with latitude.

CONCLUDING REMARKS

It should now be apparent that certain themes are beginning to recur. On the one hand, analyses of completely different data sets for the same organisms often give at least qualitatively similar results. For example, when considering the distribution of body sizes in chapter 5, I showed in figure 5.3 that species of extremely large and small size tend to appear repeatedly in different local communities across the North American continent, implying that they have larger geographic ranges. Now I have shown that these species of extreme size do indeed occupy large ranges, apparently because they have lower population densities and would otherwise have higher probabilities of extinction than their modal-sized relatives. The repeatability of the pattern suggests that it is not a statistical or methodological artifact, but a real feature of these organisms. On the other hand, the different analyses often offer different, but complementary, insights into the underlying mechanisms.

Furthermore, the same kinds of analyses of comparable data for different taxonomic groups of organisms often result in similar patterns. For example, in most cases in which I have been able to obtain comparable data for both birds and mammals, the patterns are qualitatively similar. This suggests that the underlying mechanistic processes are also very general. By analyzing the entire pattern of variation, I find limits that appear to reflect the operation of both absolute and relative constraints. I have tried to account for these constraints by invoking specific mechanisms. I should emphasize that these explanations, as reasonable as they may or may not seem, must be regarded as hypotheses. They need to be evaluated independently using not only new data for different organisms and geographic regions, but also completely different kinds of data that will permit stronger inferences to be made about mechanisms.

Most of the macroecological phenomena considered in this and the preceding chapter are interrelated. Certainly this is true of the three easily measured characteristics of species—body size, population density, and area of geographic range—that have been analyzed in some detail. None of these characteristics is independent of the others, but the pattern of dependence is complex. Although sometimes, for simplicity, I have repeatedly plotted some variables on the ordinate and spoken about the constraints of some variables on others, it would be misleading to suggest any straightforward cause-and-effect relationships.

Instead, it is necessary to adopt a multivariate perspective. We can think of individual body size, local population density, and area of geographic range as reflecting three important, interrelated ways that species have diversified to exploit the resources of large landmasses. The body size of an individual organism largely determines the quantities and qualities of energy and nutrition that it requires. The local population density indicates how many individuals with these requirements a particular local habitat can support. The size and configuration of the geographic range reflects the large-scale spatial dispersion of suitable habitats, and thus the breadth of requirements for abiotic and biotic conditions. These are not the only important variables, but together they reveal a great deal about the ecological opportunities and evolutionary constraints that have influenced the production and maintenance of the diverse species in major taxonomic groups.

Many of the empirical patterns presented above are so clear and so general that they beg for explanation. They must reflect the operation of fundamental natural laws. Many of these patterns have been known for decades. There has been much recent progress, but there are still almost unlimited opportunities to search for additional empirical patterns, to evaluate mechanistic hypotheses for existing patterns, and to develop mathematical models of these phenomena.

7

Mechanisms

Structure and Function of Individuals

In the normal course of science, discovery of a pattern at one level of organization through inductive research programs leads to the search for the causal mechanisms at lower levels of organization through deductive research programs. The pattern challenges us to develop and test hypotheses about underlying processes. I have already shown examples of this. In particular, I have hypothesized that the morphology, physiology, and behavior of individual organisms play major roles in causing, or at least constraining, large-scale patterns of distribution and abundance, both within and among species.

CONSTRAINTS ON INDIVIDUAL ORGANISMS

Additional Effects of Body Size

In the previous two chapters I showed how macroecological statistical patterns of species diversity, population density, and geographic range size are related to body size. Here I explore the mechanistic basis of these patterns by considering how size affects use of food and space, leading to dietary and habitat specialization.

Morphologists and physiologists have studied the correlates and consequences of body size for decades. There are several important books on allometry (e.g., Thompson 1917; McMahon 1973; Schmidt-Nielsen 1984) and its ecological and evolutionary implications (e.g., Peters 1983; Calder 1984; Bonner 1988; Reiss 1989). Allometry offers a valuable perspective on scaling of structure and function from which to explore further the relationship between processes at the level of individual organisms and emergent patterns at higher levels of biological organization.

Because of allometric constraints on metabolism and digestion, smaller organisms require lower rates of food consumption per individual but higher rates of energy intake per unit body mass and a higher-quality diet than their larger relatives (see chap. 5). Over a wide range of body sizes, from the largest species down to about the modal body size, these allometric constraints appear to have at least four macroecological consequences. First, species of small size must be dietary specialists in the sense that they can use only a subset of the foods that can be consumed by their larger relatives. This is demonstrated both by physiological studies of digestion and by ecological studies of diets (e.g., Wilson 1975; Townsend and Calow 1981; Owen-Smith 1988; Karasov 1990). While small species are restricted to the foods whose energy and nutrients are more easily assimilated, they can meet their lower absolute requirements by eating foods that come in small, relatively widely dispersed packages. Thus, while the smallest birds and mammals cannot graze or browse or consume entire vertebrate carcasses, they can use highly nutritious insects, seeds, and nectar that their larger relatives cannot. One might expect that small species might specialize on different kinds of foods, and that the resulting subdivision of resources might play a major role in promoting the high diversity of species in the modal size classes. This does occur to some extent, and the smaller body size classes do include representatives of guilds with specialized diets, such as insectivores, granivores, and nectarivores. The fact that local communities of mammals contain about the same number of species of modal body size as of large size (see fig. 5.3), however, suggests that the extent of partitioning of food resource types among locally coexisting species is limited. Dietary specialization may still play a major role, however, if it is associated with habitat selection.

Note, however, that the requirement for a high-quality diet need not necessarily result in specialization for particular food types (such as insects, leaves, or seeds) or for particular species or nutritional classes (such as seeds with high oil content) of those prey. Specialists can be omnivores that eat many kinds of food, so long as all items in the diet have a sufficiently high concentration of readily digestible energy and nutrients. Natural history information on the diets of many small mammals and birds supports this view; for example, in North America two of the most species-rich and abundant groups of mammals and birds are, respectively, murid rodents (e.g., deer mice) and emberizine finches (e.g., sparrows), both of which consume a variety of high-quality foods, including both seeds and insects.

Second, much of the specialization of small species is actually for habitat type rather than for food type per se. Nevertheless, constraints on the diet can cause species to be restricted to habitats where certain kinds of foods are abundant and can be harvested economically and with low risk of predation. These constraints may be caused by and reflected in differences among species in morphology, physiology, and behavior. This may be one of the reasons why some of the clearest patterns of morphology suggestive of resource partitioning are found in regional assemblages of species rather than local communities (e.g., Dayan, Simberloff et al. 1989, 1990, 1992; Dayan, Tchernov et al. 1989; Dayan and Simberloff 1994); much of the fine resource partitioning may occur in conjunction with habitat selection. Modal size does appear to be correlated with a high degree of habitat specificity, and this is reflected in the tendency of species around the modal size to have smaller geographic ranges, and even within their ranges to turn over more rapidly between habitats than their larger or smaller relatives (fig. 7.1).

Third, despite their requirements for high-quality food, modal-sized species tend to maintain population densities that are as high or higher than those of their larger relatives. As indicated above, this can be attributed at least in part to the trade-off between size and abundance and the ability of small organisms to use high-quality foods that come in small,

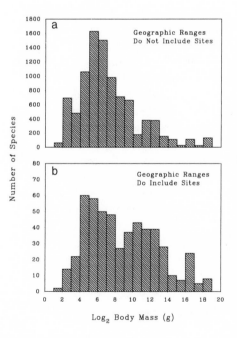

FIGURE 7.1. The cumulative frequency distributions of the body sizes of the terrestrial North American mammal species that do *not* occur in the 23 small patches of habitat analyzed by Brown and Nicoletto (see also fig. 5.3): (*a*) species whose geographic ranges do not include the sites; (*b*) species whose geographic ranges include the sites and which occur in other kinds of habitat in the immediate vicinity of the sites. To make these graphs, we first compiled the frequency distribution of body masses of all species in the continental species pool that did not occur at each site (or whose geographic ranges did not include the site), and then we summed these distributions for all 23 sites. Note that species in both categories have strongly modal size distributions with peaks at approximately 100 g. I take these distributions to indicate that modal-sized species tend to be habitat specialists, and thus restricted to small geographic ranges and to a subset of the habitats within their ranges. (Adapted from Brown and Nicoletto 1991.)

dispersed packets. Their substantial population sizes enable many small-sized species to resist extinction, even though they are restricted to a narrower range of dietary items, fewer habitat types, and smaller geographic ranges than their larger relatives. This in turn contributes to the high species diversity in the modal size classes. A small minority of modal-sized species have much higher population densities and just as large geographic ranges as their larger relatives. These species should be extremely resistant to extinction.

Finally, concomitant with the high population densities of modal-sized species are reduced requirements of individuals for space. Thus, home range or territory size (A) of both birds and mammals scales as

$$A = C_1 M^{1.0} \tag{7.1}$$

(Schoener 1968; Harestad and Bunnell 1979; Peters 1983).

This equation has some interesting implications. Since maximum population density scales as $M^{-0.75}$, requirements of individuals for space appear to increase more rapidly with size than does the space available. This problem might be resolved in two ways (fig. 7.2). Damuth (1981; see also Calder 1984) has suggested that there is a tendency for greater overlap in the home ranges of larger birds and mammals. Supporting this is the fact that many of the largest mammals are grazing herbivores, many of which not only are not strongly territorial but associate in herds or other social groups. On the other hand, I know of no evidence that the herd as a whole actually requires less space than its individuals would if they were territorial. Furthermore, many of the largest carnivorous mammals are strongly territorial (e.g., all of the large cats except lions), and flocking in birds does not appear to be associated with large body size. Alternatively, as species decrease in size, they might tend to use the environment in a more patchy way, leaving increasing amounts of space unused. This is especially likely for the less abundant species in the smaller size classes. Remember from chapter 5 (fig. 5.7) that while the population density of the most abundant species in each size class decreases as $M^{-0.75}$, there are several orders of magnitude variation in local population density among species in the small size classes. Species of small size and low average abundance must either have much larger home ranges than predicted by equation (7.1) or their home ranges must be patchily distributed, occupying only a small proportion of the available space.

New Mechanistic Hypotheses from Macroscopic Patterns

So far, I have taken a bottom-up perspective, applying insights from allometry to make connections between what is known about the structure

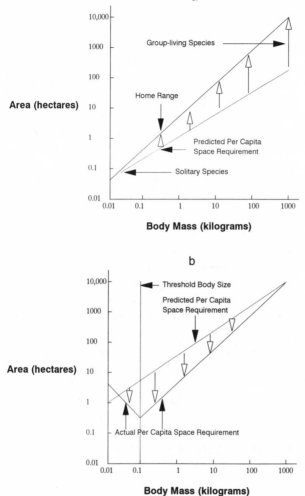

FIGURE 7.2. Two graphical models that suggest different consequences of the fact that for individuals energy requirements scale as $M^{0.75}$ whereas territory or home range size scales as $M1.0$. *(a)* Calder's (1984) suggestion that the more rapid change in home range size (solid line) than predicted on the basis of per capita resource requirements (dashed line) allows larger mammals to overlap their home ranges to a greater extent. This implies that the largest species will form herds or packs comprising multiple individuals that share large, completely overlapping home ranges. *(b)* A graphical representation of Brown and Maurer's (1989) suggestion that as body size decreases, dietary and habitat specialization enables progressively smaller species actually to use smaller home ranges (solid line) than expected on the basis of their resource requirements (dashed line), but only down to some threshold body size at which home range size area attains a minimal value. As body size continues to decrease, home range size must increase to include a sufficient quantity of patchily distributed high-quality resources. In mammals, the threshold size at which the allometry of territory size changes slope is predicted to be approximately 100 g. (Figure from Calder 1984.)

and function of individual organisms and the patterns of abundance, distribution, and diversity of species. It is also possible to take a top-down approach: to make predictions and test hypotheses about individual-level traits from species-level patterns. It is surprising, but macroecological patterns suggest that body size limits individual performance in some ways that have apparently gone undetected despite decades of research in allometry and physiological ecology.

The clearest examples concern the consequences of very small size. Recall that both the number of species and the maximum population densities of both birds and mammals decrease rapidly with decreasing body size below some modal or critical size class (figs. 5.2 and 5.7). This critical size is about 30 g for birds and 100 g for mammals. Brian Maurer and I proposed the following hypothesis to account for this pattern. As body size decreases, individuals are constrained to require higher-quality food. Over most of the range of body sizes, some species in each size class are able to do this by specializing on food types and habitats in such a way that they trade off reduced size for increased density and decreased home range size, as discussed above. At some critical size threshold, however, this becomes impossible. The constraints of size on metabolism and digestion continue to require that smaller organisms ingest higher-quality food, but individuals can no longer meet these requirements by using space in the same way. Note that the allometric equations (5.1) and (7.1) imply that metabolic requirements, E, vary as $M^{0.75}$ and home range size, A, varies as $M^{1.0}$. Thus the rate at which energy and nutrients must be harvested per unit area of territory is

$$E/A = C_2 M^{-0.25}. \tag{7.2}$$

This implies that as organisms get smaller, they must be able to harvest absolutely more energy from each square meter of their territories.

We hypothesize that organisms can only do this down to some critical size class. Organisms smaller than this threshold must alter the way that they use space: rather than collecting their food from a shrinking territory, they must operate over a larger area to find and exploit the patches where food of suitably high quality is sufficiently abundant to meet their demands.

To visualize this, imagine the landscape as a topographic surface whose peaks and valleys represent high and low food concentrations (fig. 7.3). For each class of decreasing body size, the minimum threshold for resource quality increases, and a smaller fraction of the topographic surface is above the corresponding contour. Down to some critical size, individuals whose territory sizes are decreasing with their body sizes can make a living by placing their territories under the peaks and leaving the valleys uninhabited. Below some critical body size, however, no single peak is large enough for a territory that will meet the individual's requirements, but a local population can still persist if its individuals move among and use the resources of multiple isolated peaks. This means that the entire territory or home range must increase in size, even though foraging will be restricted to isolated patches within this space. Note that rates of food resource renewal are not considered explicitly here, but if the rate of renewal for a small patch is considerably slower than the rate of harvesting, then individuals will continually deplete patches by their own foraging and be forced to move to new patches.

This hypothesis can account, at least qualitatively, for the several consequences of the fact that territory size varies more steeply with body size (as $M^{1.0}$) than individual energy requirements increase (as $M^{0.75}$) or maximum population density decreases (as $M^{-0.75}$). From the largest sizes down to the modal size, territory size can vary as $M^{1.0}$, territories can be fitted under resource peaks, and density will vary depending on the number of resource types that the species can use: the fewer resource types, whether these be kinds of foods or microhabitats required for economical foraging with acceptably low risk of predation, the more widely spaced the territories and the lower the population density for species of a given body size (fig. 7.3).

The hypothesis also accounts for the critical body size below which both maximum population density and species diversity decrease with decreasing body size. Population density will decrease because the peaks of resources above the threshold quality are separated by increasingly large valleys. Species diversity will decrease because fewer resource types will meet the quality requirements of the smallest species. Many of the consequences of the hypothesis are in effect predictions. They could be tested if sufficient data were available about such things as the spacing of territories and the numbers of resource types used by species of similar and different sizes. But such data are not generally available, and they will require some effort to obtain.

One clear prediction is made, however, that should be quite straightforward to test: below the critical body size, long-term territory size should vary inversely with body size. This prediction probably seems counterintuitive. Certainly, it suggests that the smallest birds and mammals do not conform to equation (7.1), which was fitted to data for a wide spectrum of body sizes. The prediction comes from a hypothesis developed primarily to account for the data on population densities of birds (fig. 5.7a,b), but the only data that I know of that would be sufficient to

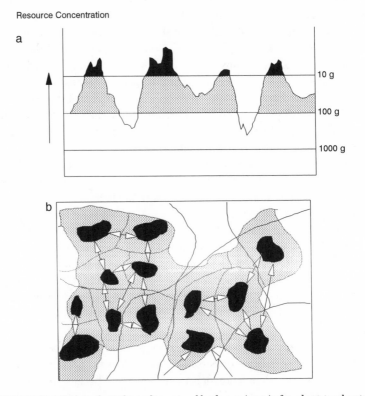

Resource Concentration

a

10 g

100 g

1000 g

b

FIGURE 7.3. (a) A hypothetical one-dimensional landscape (x axis) of productivity, showing progressively higher quality food resources concentrated in more widely separated peaks. Superimposed on this landscape are the food quality thresholds for mammals or birds (they would be scaled differently) of three different body sizes, showing that species of progressively smaller size require food of increasing quality. (b) A schematic representation of how the home ranges of the three different-sized species are distributed on a two-dimensional landscape in order to meet their requirements for energy and nutrients. The shading indicates the areas having resource levels above the thresholds for the three different-sized species in (a); the lines, boundaries of home ranges of individuals of the three species. The lines with arrows indicate that each individual of the smallest-sized species includes multiple patches within its home range. Note that individuals of the largest species are able to obtain sufficient resources by using the landscape in a continuous, fine-grained manner. Individuals of the intermediate-sized species can also meet their requirements by using space in a similar way, with the exception that some patches may contain insufficient resources to be included in their home ranges. Individuals of the smallest species must use space in a coarse-grained way, moving among multiple resource peaks and thus increasing the total area within their apparent home ranges.

test the hypothesis come from mammals. What is needed are estimates of long-term territory size or some comparable measure of space use for a number of species of varying body size below the modal size class, and these species should all inhabit the same environment so that they are subject to the same constraints of resource availability. Data on use of space by the eleven most abundant species of desert rodents at my experimental study site meet these criteria (Brown and Zeng 1989). The median lifetime movement of these species was indeed inversely correlated with their body sizes (fig. 7.4b), especially for the ten species that weighed less than 100 g. Note, too, that while population densities of these rodents were highly variable, the highest values were exhibited not by the smallest species, but by ones much nearer the modal size for mammals (fig. 7.4a).

A second prediction, but this time a macroscopic one, is that the minimum size of geographic ranges should increase with decreasing body size below the modal size. The reason is straightforward. If the minimum size of geographic range is a relative constraint, set by increasing probability of extinction with decreasing population size, and if population density decreases in species smaller as well as larger than the modal size, then the minimum area of geographic range should increase as body size departs from the modal size. There is a hint of this in the data for North American birds and mammals: the smallest species in both taxa appear to have larger ranges than their somewhat larger relatives (see fig. 6.3). It is difficult to claim that this trend is statistically significant, however, because the fewer species in the smallest size categories reduce the probability that any one of them will have an extremely low value for range size.

My student Pablo Marquet has devised a better way to test this prediction. He reasoned that if modal-sized species could maintain the highest population densities and the smallest geographic ranges, then species of approximately modal size should be the only ones inhabiting the smallest islands, and the range of body sizes should increase with increasing size and species richness of islands. Marquet and M. L. Taper (unpub.) compiled and analyzed a large data set on the body sizes of mammals occurring on oceanic islands throughout the world. The pattern (refer back to fig. 5.6b) strongly supports the prediction: the range of body sizes is small and the species are near the modal size when species richness is low, and the range of sizes increases in a very regular pattern with increasing species richness. Further, Marquet and Taper performed a randomization test that showed that this pattern cannot be explained by a null hypothesis of random sampling of insular faunas from a continental or global species pool.

Both the model of body size—dependent use of space and the model of optimal body size presented in chapter 5 predict the pattern of body

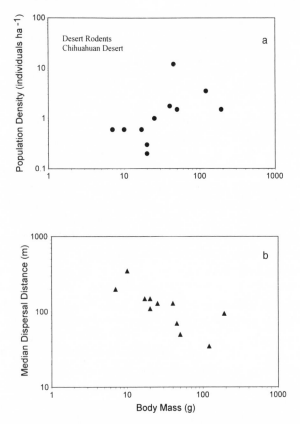

FIGURE 7.4. (*a*) Population densities and (*b*) lifetime movements as a function of body mass for the eleven common species of rodents on my experimental study site in the Chihuahuan Desert. Note that the highest populations are attained by species weighing between 50 and 150 g, as might be expected from theory and from other data on terrestrial mammals (see chap. 5 and fig. 5.7). Note also that lifetime movements, which I take as a good measure of long-term home range sizes, are inversely correlated with body sizes for species weighing less than 100 g. This is not expected from the standard allometric equations, but is predicted by the theory developed in this chapter and in chapter 5. (Figure adapted from figure 4 in J. H. Brown and B. A. Maurer, "Macroecology: The Division of Food and Space among Species on Continents," *Science* [1989] 243: 1145–50. (c) 1989 AAAS.)

sizes on continents and islands. While these two models invoke somewhat different proximate mechanisms, their converging predictions are not independent. They both reflect increasingly severe energetic constraints on individuals as their body sizes decrease below some critical value. The optimal size model also predicts changes in trends in life history traits as body size departs from the optimal size: for example, clutch size should decrease, while interval between clutches, age to first reproduction, and life span should increase, so that the intrinsic rate of natural increase (r) should also decrease (Brown, Marquet, and Taper 1993).

This prediction leads to a possible explanation for a pattern that has intrigued me. Some of the fastest rates of evolutionary change that have been measured are the dwarfing trends that occurred in some populations of large mammals that were isolated on continental islands by rising sea levels at the end of the Pleistocene. Elephants on Sicily and Malta decreased by an order of magnitude or more in body mass in only a few thousand years (Roth 1990; see also fig. 5.6a and Lister 1989 on deer)! I could understand why small, isolated populations of such large mammals might go extinct, but what kind of strong natural selection acting differentially on individuals within populations could cause the decrease in body size and lead indirectly to increased population size and decreased probability of extinction?

The optimal size model of Brown et al. (1993) suggests a hypothesis. Let's assume that a population of a large herbivore that survived the initial isolation of the island found itself in an environment with fewer competitors of smaller size (because at least some of the species in these smaller size categories did go extinct). The megaherbivore could exploit resources that the missing competitors had been using by evolving a smaller size and a more specialized, higher-quality diet. But why would there be strong selection to do so, especially considering that there appears to be no more usable energy available to small organisms than to large ones (chap. 5 and figs. 5.9 and 5.10)? I suggest that the answer lies in the inherent fitness advantage that comes from being nearer the optimal size. The reduced time to first reproduction and shorter intervals between litters, and the possible increase in litter size, that would come as correlates of reduced body size would enable smaller phenotypes to produce more offspring more rapidly than their larger relatives. Thus, according to this hypothesis, reduced competition from smaller-sized competitors does not by itself cause the selection for dwarfing of large herbivores isolated on small islands. Instead, the reduced competition provides the ecological opportunity for smaller-sized phenotypes to realize the inherent life history and fitness advantages of being closer to the optimal size. A similar argument would hypothetically explain microevolutionary trends toward insular gigantism in the smallest mammals (see fig. 5.6a and Lomolino 1985).

Thus, macroecological patterns and resulting theoretical explorations (both above and in chapter 5) have led us to make testable predictions that many ecological variables do not scale monotonically with body size, as would be expected from allometric equations. More data and analyses will be required to make definitive tests of these hypotheses. In the case of territory size and life history variables, the published allometric equations (e.g., Peters 1983; Calder 1984) have been derived empirically by compiling all available data for a taxon, such as birds or mammals, and then determining the best fit power function. Because the data typically span many orders of magnitude of body mass (e.g., seven orders of magnitude for land mammals, eight if the whales are included; five for flying birds, six if the nonvolant ratites are included) but only one to two orders of magnitude separate the smallest from the modal-sized species, and because with ecological parameters (in contrast to many morphological and physiological variables) there is inevitably a large proportion of the variation that is not accounted for by the best fit allometric regression equation, it is perhaps not surprising that systematic deviations in the smallest species have not yet been detected.

There is at least anecdotal evidence to support these predictions. For example, maximum life span and clutch size are supposed to scale positively and negatively, respectively, with respect to body size in both birds and mammals (e.g., Peters 1983; Calder 1984; Harvey and Nee 1991; Charnov 1992). But some of the smallest representatives of these taxa are conspicuously long-lived: the 3.5 g broad-tailed hummingbird (*Selasphorus platycercus*) is known to live at least 12 years in the wild (Calder 1989), and several species of similarly sized insectivorous bats (family Vespertilionidae) have lived more than 10 years (Findley 1993). Hummingbirds and the smallest bats also have very low clutch and litter sizes, producing only two young per year.

While such isolated cases are tantalizing, they are inadequate to evaluate the above hypotheses. To do so will require high-quality ecological data on the smaller representatives of taxonomic groups and analyses that can detect changes in sign or slope below some critical body size if they occur. When the results are in, it will be interesting to see which, if any, traits of the smallest organisms deviate in the way I have predicted from the allometric trends exhibited by their larger relatives. And even if the patterns that I have predicted are corroborated, some important conceptual links still need to be made. Brown et al. (1993) suggest that a size of about 100 g in mammals (and presumably about 30 g in birds) represents a fitness optimum set by energetic constraints. From this we have argued that components of fitness, such as clutch size, life span, and r, should exhibit maximal or minimal values at this size. It is far from clear, however, why ecological characteristics such as home range size, population density, area of geographic range, and species richness should also attain their extreme values at this same size. There is room for much more work here.

DIFFERENCES BETWEEN BIRDS AND MAMMALS

The Pattern: Differences in Size, Population Density, and Energy Use

When comparable data for North American birds and mammals are available, the macroecological patterns of the two groups are often qualitatively similar. They are not quantitatively similar. Although the smallest birds and mammals are of nearly identical size (approximately 2 g), the largest terrestrial mammals are about two orders of magnitude larger than the largest land birds (see fig. 5.2). Exploring the consequences of this difference raises some interesting questions about population densities and rates of energy use and leads to testable hypotheses about individual-level characteristics.

What if population sizes or rates of population energy use as a function of body mass for both mammals and birds were plotted on the same graph? Initially I was unable to do this, because I had good census data over many local sites only for a large sample of bird species. I did, however, predict how the relationships for mammals should differ by making four assumptions: (1) the constraint envelopes for mammals and birds have the same shapes (i.e., the boundary lines have approximately the same slopes, as shown in figure 7.5; see also figures 5.7b,c); (2) the constraint envelopes for mammals differ from those for birds in size and position because of the orders of magnitude larger range of mammalian body sizes; (3) the population densities of the largest mammals are not significantly lower than those of the largest birds (this critical assumption requires that bears, cougars, moose, and elk be no less dense on average than eagles, condors, turkeys, and grouse, which does not seem unreasonable); and, for estimating rates of energy use only, (4) the average field metabolic rates of birds and mammals scale similarly with respect to body mass (for the appropriate equations see Peters 1983; Calder 1984).

These assumptions predict the constraint envelopes for birds and mammals shown in figure 7.5. It is apparent immediately that mammals are predicted to have maximum population densities and rates of energy use that exceed those of birds of comparable size by nearly two orders of magnitude. Is this correct? If so, such a large difference between two such well studied taxa might be common knowledge. In fact, I am not aware of any published studies that call attention to differences in the population densities of birds and mammals. A few studies do suggest that birds are quantitatively much less important consumers of energy than mammals in some ecosystems (e.g., Wiens 1973; Wiens and Dyer 1977).

There is, of course, a large literature containing data on the densities of local populations of both birds and mammals. First Pablo Marquet and

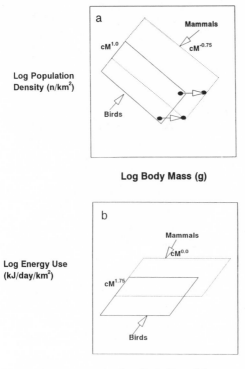

FIGURE 7.5. Hypothesized constraint spaces for (a) population density and (b) population energy use as a function of body mass in land birds and mammals. Since the largest mammals are two orders of magnitude larger (arrows) than the largest birds but the population densities of the two taxa seem to be similar, we predicted that within a size category the most abundant mammals would be approximately two orders of magnitude higher in population density than the most abundant birds. Since the rate of energy use (metabolism) is very similar for birds and mammals, the above difference in population density should also be reflected in about two orders of magnitude difference in population energy use.

then Marina Silva (M. Silva and J. H. Brown, unpub.) compiled and analyzed a large sample of these data to test the predictions. Marquet's preliminary results clearly suggest that maximum population densities are indeed about two orders of magnitude higher for mammals than for birds of comparable size (fig. 7.6). The predicted differences in both population density and energy use are clearly seen in figures 5.7b,c and 5.9, in which the complete data compiled by Silva are plotted.

A Hypothesis: The Constraints of Flight

This macroscopic pattern, a difference of such magnitude between birds and mammals (which are so similar in many other respects), calls for a mechanistic hypothesis. What constraint on individual-level function causes at least an order of magnitude difference in population density and energy use? The obvious candidate is some characteristic(s) related to flight, since all of the North American land birds are capable of flight and bats, the flying mammals, were excluded from the terrestrial mammal data set. But how does flight affect population density?

I suggest the following hypothesis. The aerodynamic constraints of

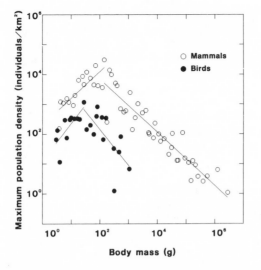

FIGURE 7.6. Distribution of maximal population density as a function of body mass in terrestrial birds and mammals. The values for the species of terrestrial mammals (open circles) and land birds (solid circles) that represent the highest population densities in these groups are plotted. Separate regression equations have been fitted for the mammal species weighing more and less than 100 g, and similarly for bird species weighing more and less than 30 g. Note that, as predicted (see fig. 7.5 and text), the most abundant mammals have population densities more than an order of magnitude higher than those of the most abundant birds of similar size. Also, maximal density reaches a peak value at approximately 100 g in mammals and 30 g in birds. (From P. A. Marquet, unpub.)

flight place severe limits on the shape of the organism and the body mass that can be lifted and transported. In particular, a flying bird is limited in the amount of space and mass that can be devoted to food processing and digestion. The size (and length) of the digestive tract is smaller and the amount of food contents being processed is less in birds than in mammals. As a consequence mammals can keep food in the gut for a longer time, extract a larger proportion of the energy and nutrients, and thus subsist on lower-quality foods. This in turn enables mammals to use a larger proportion of available resources and attain considerably higher population densities than birds of the same size and dietary guild.

Like most good hypotheses, this one has logical consequences that lead to additional, testable predictions. Some of them are the following:

1. Mammals have longer digestive tracts, larger gut capacities, and slower passage times than birds of similar size and diet. It is hard to find data on the last two variables to make the appropriate comparisons, but reasonably good data on allometric scaling suggest that mammals do indeed have longer absorptive guts than birds (W. Karasov, pers. comm.).

2. If birds and mammals of the same size consume the same foods, then

mammals should be able to extract a larger proportion of the energy and nutrients. There are few data that can be used to test this prediction, because rarely is diet controlled in a way that makes comparison valid. Nevertheless, there are a few suggestions that mammals do have higher extraction efficiencies than birds (W. Karasov, pers. comm.).

3. The opposite side of the previous prediction is that mammals can maintain positive energy and nutritional balance while feeding on lower-quality foods. This prediction is supported by several qualitative observations. Leaves, twigs, and bark, which contain high concentrations of fiber (cellulose and lignin), are especially difficult for vertebrates to digest. There are numerous species of folivorous mammals spanning a wide range of body sizes, and some of them include large quantities of twigs and bark in their diets. Very few birds are largely or exclusively folivorous, and those that feed on vegetation (e.g., hoatzins, grouse, geese) tend to consume plant parts, such as fresh leaves, buds, and storage roots, that are relatively high in energy and nutrients and low in fiber (e.g., Morton 1978; Grajal et al. 1989). Folivorous birds also tend to be heavy-bodied and relatively poor fliers.

4. Birds that have become secondarily flightless should be able to escape at least some of the constraints limiting flying birds. In particular they should be able to consume food types and attain population densities that are more similar to those of mammals. This prediction is supported by the fact that flightless birds endemic to oceanic islands are often described as filling mammalian niches and by the few data suggesting that some of these bird species have higher population densities than their volant relatives (M. Silva and J. H. Brown, unpub.). My hypothesis also predicts that the large, flightless land birds found on continents other than North America (ostriches, cassowaries, emus, and rheas) should attain higher population densities than would be predicted by extrapolating from smaller, flying birds (although not necessarily as high as those of some mammals of comparable size, because of lingering evolutionary constraints). Again, the limited data appear to support this prediction (e.g., value for ostrich in fig. 5.7b).

5. The opposite side of this prediction is that the aerodynamic and digestive constraints on bats and birds should be similar, and consequently bats should be more similar to birds than to nonvolant mammals in diet, population density, and energy use. The virtual absence of folivory among bats is consistent with this prediction. To evaluate the prediction about population densities of bats, it is necessary to use measures of "ecological density" (sensu Damuth 1981, 1987). Most bats roost in colonies containing tens to millions of individuals, but they disperse widely to feed. The very limited data on densities of such foraging bats suggests that they are comparable to those of birds and considerably lower than those of nonvolant mammals—and the same is true for population energy use (figs. 5.7b,c and 5.9).

CONCLUDING REMARKS

This chapter explores some of the connections between the kinds of individual-level traits studied by functional morphologists and comparative physiologists and the species-level traits revealed by the macroecological approach. It is clear that the study of these relationships can benefit from both top-down and bottom-up approaches.

On the one hand, studies of the structure and function of individual organisms provide invaluable insights into important mechanisms that may cause ecological, biogeographic, and macroevolutionary patterns. I have emphasized only the kinds of insights that can come from studies of allometry in birds and mammals.[1] Many other kinds of connections could be made. For example, the powerful and pervasive influence of evolutionary constraints is just beginning to be fully appreciated. Now that it is possible to reconstruct phylogenetic lineages with considerable accuracy, it should be possible to consider the influence of evolutionary constraints on ecological, biogeographic, and macroevolutionary patterns in much the same way that I have dealt with the effects of allometric constraints (e.g., see Lessa and Patton 1989; Taylor and Gotelli, in press). Here, too, many of the species-level attributes may depend on structural and functional limits at the level of individuals.

On the other hand, the macroecological approach can offer potentially valuable insights into lower levels of biological organization. Constraints that show clearly as species-level patterns may suggest limits on the structure and function of individual organisms that have not yet been appreciated. Thus, my efforts to explain statistical patterns in the abundance and distribution of many species have led to the development and preliminary evaluation of hypotheses about the morphology and physiology of individual organisms.

Making these connections between the structure and function of individual organisms and their ecological, biogeographic, and evolutionary consequences requires the synthesis of data, theory, and analytical methods across the different disciplines that have specialized on the different levels of biological organization. My efforts at synthesis have focused on the common themes of energetics and allometry. This leads me to make explicitly a point made implicitly in chapter 5: a major impediment to synthesis and communication is the use of different currencies—rates of energy or material use as opposed to rates of change in numbers of indi-

---

1. Farlow (1993) has recently applied concepts of allometry and energetics to make some very interesting speculations about the ecology and biogeography of the large carnivorous dinosaurs.

viduals, genotypes, or alleles. I will return to this problem in chapter 10, where I will suggest that the long-term solution is to develop a common energetic/thermodynamic currency. In the meantime, we must make an effort to communicate across the disciplinary barriers: reading the literature, understanding the perspective, and applying the ideas and data. Those who fail to do at least this much will miss unique opportunities to contribute to the synthesis of modern biology.

8

Mechanisms
Population Dynamics and Interspecific Interactions

So far, much of the emphasis in this book has been on mechanisms at the level of individuals that operate as constraints to set the boundaries of macroecological patterns. Much less has been said about the processes that influence population structure and dynamics within the limits set by these constraints. In part this is because, in making hypotheses about absolute constraints, it has often been possible to skip from species-level patterns to individual-level processes without considering in detail the intervening population-level phenomena. Thus, for example, if it is impossible for an endothermic bird or mammal smaller than about 2 g to function as an organism, it is moot to consider what kinds of population structure and dynamics such an animal might exhibit.

A GENERAL COMMENT ON MECHANISMS

Mechanisms at Individual and Population Levels

Within the boundaries set by the individual-level constraints, however, the species-level patterns must be attributed in large part to population-level processes. Where individuals live and what abundances they attain are determined by their interactions with one another, with their abiotic environment, and with other organisms. This was recognized implicitly in the previous chapter by the suggestion that there are relative as well as absolute constraints. The relative constraints were hypothesized to reflect the influence of probabilistic population processes.

On the one hand, macroecological patterns cannot be explained without knowing how populations of a species respond to spatial and temporal variation in the environment. Chapters 3 and 4 pointed out that abundance and distribution reflect the outcome of the interaction between environmental limits and the intrinsic capacity of populations to increase and to disperse. Of course, population processes are influenced by characteristics of individual organisms that affect stress tolerance, reproductive effort, dispersal, foraging, predator avoidance, and so on. One such characteristic is body size, but there are many others.

On the other hand, the premise of macroecology is that it is possible to learn a good deal about the emergent properties of entire biotas without having to know many of the characteristics of their component species, populations, and individuals. The goal is to explain macroscopic patterns in terms of general processes, rather than specific details. I have done this at the individual level by exploring the use of allometric relationships to make and test hypotheses about constraints of body size and energetics.

Some parts of the picture are still missing. For example, it would be nice if morphologists and physiologists could explain the lower limit on body mass in birds and mammals. Nevertheless, the answer would probably be of limited generality, because the specific mechanism that sets the limit at approximately 2 g in both birds and mammals is not the same as the one that sets the limit at smaller sizes in reptiles, fishes, and insects. Furthermore, much progress can be made without knowing the exact, taxon-specific answer. The phenomenon of interest is that individual-level constraints appear to set a sharp, nearly absolute limit on the minimum size of each taxonomic group.

Can similar phenomenological explanations be made in terms of processes at the population level? The exercise of predicting extinctions of mountaintop mammal populations in response to global climate change (chap. 1) suggests that they can. This example may be misleading, however. Islands and other isolated habitats are relatively simple systems. Ecologists, biogeographers, and evolutionary biologists have often treated insular systems as empirical models and assumed that the same processes that determine their structure and dynamics also operate and have similar effects in more complex continental systems.

The Spatial Pattern of Population Dynamics

Chapter 1 describes the predictions that Kelly McDonald and I (1992) made about changes in the composition of the small mammal faunas of isolated mountaintops in the Great Basin that would occur in response to global warming. It was possible to make these predictions because of three patterns of correlation. First, the important environmental changes were assumed to be correlated over the entire region; it was assumed that a global increase in temperature would cause an equal upward shift in the climatic and vegetation zones on all mountain ranges in the region. Second, the responses were assumed to be correlated among different populations of the same species on different mountains, because all populations would experience similar environmental changes. Third, the responses were also assumed to be correlated among different species on the same mountains (and also across different mountain ranges), because the species share requirements for cool, mesic, high-elevation habitats. It was assumed that all populations of all species would contract their ranges and decrease their populations in proportion to the reduction in habitable area. I could potentially make similar predictions about changes in the biotas of an oceanic archipelago that would occur in response to rising sea level, because the reduction of habitable area could be assumed to have correlated effects on all populations of all terrestrial species across all islands.

These kinds of correlations in both environmental variation and population response provide a potentially powerful basis for making inferences about the general processes that underlie macroecological patterns. But do these kinds of correlation structure occur in continental situations, as they apparently do on islands? Several considerations suggest that temporal fluctuations of populations of the same species should be correlated over space; that is, the time series for nearby populations should be strongly positively cross-correlated, and the degree of this correlation should decline with increasing spatial separation between populations. I can give three reasons: (1) important environmental variables tend to be spatially autocorrelated, so that nearby environments should experience similar temporal fluctuations; (2) these spatially autocorrelated environmental variables should affect nearby populations of the same species in similar ways; and (3) dispersal of individuals among populations, which should be in inverse proportion to the distance separating them, should also tend to cause spatially autocorrelated dynamics. The first point, which is essentially the same as the assumption of spatial autocorrelation in environmental variables made in chapter 4, must be true, at least at small spatial scales for certain environmental variables (e.g., characteristics of climate and soil).

However, it is possible to imagine exceptions to the last two assumptions. Consider some hypothetical species of bird or mammal living in a region of highly variable precipitation. It is not at all hard to imagine populations in two adjacent habitats showing opposite fluctuations, because the habitat that was most favorable during the drought could become too wet during the wet year, and vice versa. Similarly, a species that is dependent on a specific kind of vegetation could easily show opposite year-to-year population trends in nearby habitat patches in different stages of succession: populations would be increasing in some patches as they entered a favorable stage, but decreasing in other patches as more advanced succession made them unsuitable. In both of these cases, dispersal probably would contribute to the negative correlation. Individuals would tend to move from the less favorable to the more favorable environments, increasing the opposite population trends due to differences in birth and death rates.

Thus, it is not at all clear that there should be a general tendency of spatial autocorrelation in population dynamics within a species over a complex continental landscape. What about between species? Similar considerations apply, even for species that are members of the same taxonomic group or ecological guild. While their similarities may sometimes cause them to respond similarly, even small differences can be sufficient to cause negative correlations. It is difficult to make a priori predictions without also making additional highly restrictive assumptions. The question of spatial patterning of population dynamics within and among species thus becomes primarily an empirical issue.

A CASE STUDY: INDIVIDUALISTIC TEMPORAL AND SPATIAL DYNAMICS OF BREEDING BIRD POPULATIONS

The BBS Data Set

In collaboration with Mark Taper and Katrin Bohning-Gaese, I have been analyzing the spatial and temporal fluctuations of North American breeding land bird populations using the Breeding Bird Survey. As pointed out in chapter 4, the BBS is a unique and potentially invaluable data set because of its combination of spatial and temporal coverage (e.g., Bystrak 1979; Robbins, Bystrak, and Geissler 1986). It and the somewhat similar Christmas Bird Counts provide virtually the only long-term census data covering a large geographic area for any vertebrate animals in North America. So far, most of our analyses have been restricted to a group of small insectivorous passerines: 59 species of warblers, vireos, chickadees, wrens, and a few other families.

Individualistic Spatial and Temporal Dynamics

One question, then, is to what extent the year-to-year fluctuations of populations of the same species are correlated over space. The answer is that there are significant positive correlations only on relatively small spatial scales, such as within the same BBS "stratum," a biome or region of relatively similar climate and vegetation, or for sites separated by relatively short distances (see fig. 4.8). At larger scales, such as among different strata, the positive correlations are barely more significant than the 2.5% that would be expected by chance (table 8.1; K. Bohning-Gaese, M. L.

Taper, and J. H. Brown, unpub.). These results should be interpreted with caution because there is considerable sample error in individual BBS censuses that may obscure some spatial patterns (see Bystrak 1979; Robbins, Bystrak, and Geissler 1986). The data are sufficient, however, to reveal many large-scale trends: there are highly significant decades-long, stratum-wide increasing and decreasing trends in the populations of many species (table 8.1). This is apparent in figure 8.1, in which the red-eyed vireo (*Vireo olivaceus*) shows statistically significant increasing, decreasing, and curvilinear trends in different strata. Therefore, the absence of strong correlations in year-to-year fluctuations among nearby sites, even in abundant species for which sampling errors should be minimal, is especially noteworthy.

A second question is what spatial and temporal patterns are found among different closely related and/or ecologically similar species. Within the same stratum, different species exhibit widely varying patterns of temporal fluctuation (fig. 8.2; K. Bohning-Gaese, M. L. Taper, and J. H. Brown, unpub.). Of the 59 species of small insectivores with sufficient data for analysis, nearly all increased significantly in some strata and decreased in others, and each of the 22 strata had some species that increased significantly and others that decreased (table 8.2). The general answer, then, is that each species tends to be highly individualistic, increasing in some regions and decreasing in others, with seemingly little relationship to the dynamics of other species.

These results warrant several comments. First, the species studied include long-distance migrants that winter in the Neotropics, short-distance

TABLE 8.1. Variation in the temporal patterns of population fluctuation of single species of small insectivorous passerine birds across multiple regions (BBS strata or biome types) in the eastern and central United States over a 25-year period, 1965–1990.

	Number of species
Species that occur in two or more strata	47
Species with declines in at least one stratum and increases in at least one other stratum	36 (77%)
Species with significant declines in at least one stratum and significant increases in at least one other stratum	18 (38%)
Species with increases in all strata	7 (15%)
Species with significant increases in all strata	1 (2%)
Species with declines in all strata	4 (9%)
Species with significant declines in all strata	0 (0%)

Source: K. Bohning-Gaese, M. L. Taper, and J. H. Brown, unpub.

Note: Most of the forty-seven species increased in some regions but decreased in others, and only one species showed statistically significant increases (and none had significant decreases) across all regions.

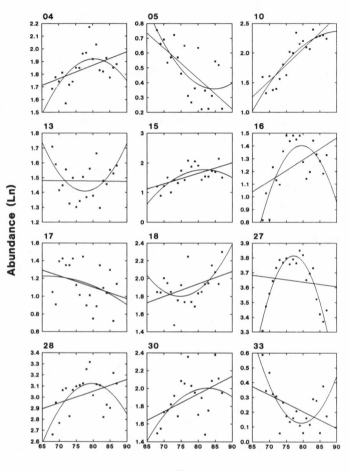

Year

FIGURE 8.1. Patterns of fluctuation in the abundance of a bird species, the red-eyed vireo (*Vireo olivaceus*) in twelve different regions (biome types) over a 25-year period, based on data from the Breeding Bird Survey (BBS). The number above each graph refers to a different biome type (BBS stratum) in the eastern and central United States; the y axis gives abundance in the logarithm (plus 1) of the average census counts for each year, and the straight and curved lines are best-fit linear and quadratic regression equations. Note the enormous variation among the regions, even in adjoining regions as indicated by consecutive numbers, in the decades-long pattern of fluctuation. This degree of variation is typical of most species of small insectivorous passerines that we have examined. (From K. Bohning-Gaese, M. L. Taper, and J. H. Brown, unpub.)

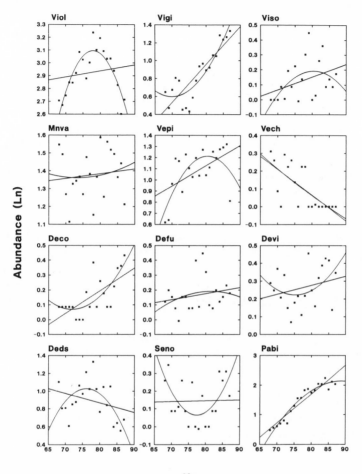

Abundance (Ln)

Year

FIGURE 8.2. Patterns of variation in the abundances of twelve species of vireos and warblers within the southern New England region (stratum 12), based on data from the Breeding Bird Survey. The four-letter codes give the first two letters of the genus and species; the data are plotted as in figure 8.1, and the straight and curved lines are best-fit linear and quadratic regression equations. Note the enormous variation among these ecologically similar species of small insectivorous passerines in their decades-long patterns of fluctuation within a relatively small region. (From K. Bohning-Gaese, M. L. Taper, and J. H. Brown, unpub.)

TABLE 8.2. Variation in the temporal patterns of population fluctuation of multiple species of small insectivorous passerine birds within a single region (BBS stratum or biome type) in the eastern and central United States over a 25-year period.

	Number of strata
Strata with declines in at least one species and increases in at least one other species	20 (91%)
Strata with significant declines in at least one species and significant increases in at least one other species	17 (77%)
Strata with increases in all species	2 (9%)
Strata with significant increases in all species	0 (0%)
Strata with declines in all species	0 (0%)
Total strata	22

Source: K. Bohning-Gaese, M. L. Taper, and J. M. Brown, unpub.

Note: The vast majority of regions contained some species that increased and others that decreased, and in none of the regions did all of the species exhibit either statistically significant increases or declines.

migrants that winter in temperate regions of the southern United States and northern Mexico, and year-round residents. There were no conspicuous differences in the spatial patterns of interannual population trends of migrants and residents. Thus, despite their potential for long-distance dispersal, migrants (in which adults may return year after year to the same localities to breed, while young of the year often disperse considerable distances) did not consistently show either stronger or weaker patterns of spatially correlated population fluctuations than more sedentary species.

Second, strong regional (i.e., within-stratum) correlations were notable for their absence, both within and among species. This is despite the fact that the winters of 1976–77 and 1977–78 were very severe, were supposed to have "caused some of the heaviest bird mortality ever recorded in the central and eastern states" (C. S. Robbins, pers. comm.), and thus might have caused correlated fluctuations within and among some short-distance migrant and permanent resident species. This lack of spatial correlation cannot be attributed entirely to sampling errors obscuring all patterns in the data, because there is abundant evidence of significant stratum-level population trends in individual species, but these are usually uncorrelated (or only very weakly correlated) across strata within species and among species within strata.

Third, the facts that nearly every species increased significantly in some strata and decreased in others, and that every stratum had some species that increased while others decreased, reinforce the comments about Gleasonian individualism made in chapter 3. The general pattern is a lack of consistency in the details of the local population dynamics, with the conspicuous exception that there were significantly more cases of strong increases and decreases than would be expected by chance. This suggests that even species in the same taxonomic group (genus or subfamily) or ecological guild differ enough in their niche requirements that they respond highly individualistically to spatial and temporal variation in the environment.

EMERGENT PATTERNS IN AVIAN POPULATION DYNAMICS

Effects of Climate and Food Supply on Wintering Bird Populations

The population dynamics of breeding birds revealed by the BBS do not show strong patterns of spatial correlation among populations of the same species at nearby sites or among different species at the same site. It is important to ask, however, whether this is a general result. Several studies of wintering birds suggest that they may show more clearly the effects of common environmental limiting factors. In particular, the distributions and densities of many species may be strongly influenced by weather conditions.

Terry Root (1988a,b,c) has used another continent-wide data set, the Christmas Bird Counts (CBC) coordinated by the National Audubon Society, to study several aspects of the distribution and abundance of wintering birds. One of the most interesting results is that the northern boundaries of the ranges of many species are strongly correlated with environmental temperature, and in particular with the coldest winter temperatures. For a surprisingly large number of these species, the northern range boundary appears to correspond closely to minimum temperatures that would cause a twofold increase in metabolism above the resting rate for thermoregulation. This, together with somewhat more anecdotal evidence of widespread mortality and well-documented southward shifts in the distributions of the northernmost populations in severe winters (Bock and Lepthien 1972, 1976; Bock 1982; Robbins et al. 1989), has led Root to suggest that the northern limits are set by the capacity to tolerate physical stress.

Root's studies are important, but her conclusions about northern range boundaries should be taken with caution for two reasons. First, Repasky (1991) has severely criticized Root's analyses on biological and statistical grounds. Most of his criticisms revolve around the issue of implying causation from correlation. This is often a problem at large spatial scales where experimentation is impossible, but it means that investigators must take great care in making inferences. Second, even if the northern range boundaries of many of these species are determined largely by climatic conditions, the populations may not be limited directly by physiological tolerance of cold stress. Many small birds that winter in temperate North America have the capacity to increase metabolism manyfold to thermo-regulate at low temperatures so long as sufficient food is available (Dawson and Carey 1976). This fact raises the possibility that correlations between range boundaries and minimum temperatures reflect much more complex relationships between the climatic regime and population dynamics.

Several other studies that use CBC data on abundances and distributions of wintering birds support this suggestion. Laurance and Yensen (1985) analyzed year-to-year fluctuations in the densities of sparrows wintering in the climatically severe Great Basin region. They found that abundances of most species were consistently negatively correlated with depth of snow cover. Dunning and I (Dunning and Brown 1982) had earlier performed a similar study in southeastern Arizona, where lasting snow is infrequent. We found that abundances of most sparrow species were consistently correlated with the amount of precipitation that fell in the previous summer, a climatic variable that should index production of the seeds that constitute the primary food of wintering sparrows. Taken together, these two studies suggest that the densities, and by inference, the distributional limits (see also Pulliam and Parker 1979), of wintering sparrows are determined in large part by climatic factors that determine the availability of food.

Bock and Lepthien (1972, 1976; see also Bock 1982) have used CBC data to analyze so-called irruptions in wintering populations of North American birds. Several species, including hawk owls, snowy owls, rough-legged hawks, red-breasted nuthatches, pine siskins, common redpolls, evening grosbeaks, and red and white-winged crossbills, which normally winter in the far north, migrate much farther south in certain years. Bock and Lepthien have shown that the timing of such irruptions tends to be correlated among species and to be associated with climatic variables that should indicate reduced food supply on the normal wintering grounds.

Thus, some data sets do show consistent patterns of spatial and temporal variation in population density within and among North American bird species. These examples do not suggest the same degree of highly individualistic, species-specific variation that we found in our analyses of BBS data for small insectivorous birds on their breeding grounds. Many avian ecologists (e.g., Fretwell 1972; Pulliam and Mills 1977; Pulliam and Parker 1979; Pulliam 1983; see discussion in Wiens 1989) have suggested that the ecologies of breeding and wintering populations of migratory songbirds are very different, and that it is often much easier to find evidence of population regulation and community structure during the winter.

Effects of Predation on Breeding Populations of Small Insectivorous Birds

Having emphasized the pervasiveness of individualism in the population dynamics of breeding songbirds, I must point out that some general trends can also be found. In chapter 4 I showed that frequency distributions of the magnitude of spatial variation in abundance of bird species over all BBS census routes within their geographic ranges exhibit a very general pattern also observed in other organisms (see fig. 4.5).[1] Further, the spatial pattern of this variation appears to show some general features: abundance tends to be more similar at nearby sites than expected by chance, and it tends to decline from the center toward the edges of the geographic range.

A distinct but not so general pattern seems to be present in the long-term trends of small insectivorous songbird populations. When census data for all of eastern and central North America are combined, many of the species exhibit significantly increasing or decreasing populations, especially during the decade 1976—86 (Robbins et al. 1989; Bohning-Gaese, Taper, and Brown 1993). An analysis of these trends by Robbins et al. suggested that many of the species that decreased are Neotropical migrants. This supports Terborgh's (1989; but see Hutto 1989) contention that clearing of tropical forests has already reached a level that is causing detectable declines in some species of songbirds that breed in temperate North America but winter in the tropics.

Like Robbins et al., we also tried to identify attributes of species that are correlated with the decades-long population trends, but our analysis produced different results (Bohning-Gaese, Taper, and Brown 1993). First, while a small proportion of species, including tropical migrants, decreased, a much larger number of species, also including many tropical migrants, increased. Second, the tendency to increase or decrease was associated with several traits, including not only migratory status but also taxonomic group and nest characteristics. In fact, our analyses suggested

1. It would be interesting to know whether any similar emergent pattern occurs in temporal variation in abundance. To address this would require a very long time series of censuses at the same site taken sufficiently far apart so as not to include the trivial short-term autocorrelation. Dave Mehlman, George Stevens, and I analyzed a 10,000-year pollen record of the abundance of a pine species and found a pattern similar to the spatial ones shown in figure 4.5 (J. H. Brown, D. W. Mehlman, and G. C. Stevens, unpub.), but there is room for much additional work on this problem (see also Pimm and Redfearn 1988; Pimm 1991).

that population declines in these small insectivorous passerines were more suggestive of effects of nest predation and cowbird brood parasitism on the breeding grounds in temperate North America than of deforestation on the wintering grounds in the tropics.

Inferring cause and effect from these correlations is difficult, however, because the variables are distributed among species as suites of traits that are correlated with one another. The traits are also associated with taxonomy (at the generic and family levels) and thus apparently reflect constraints of phylogenetic history. Nest predation and brood parasitism on the breeding grounds, deforestation on the wintering grounds of species that winter in the tropics, severe weather on the wintering grounds of species that winter in temperate North America, and other as yet unmeasured factors may all affect the decades-long trends in populations of songbirds that breed in eastern North America.

The above patterns show that it is possible to sort through all of the Gleasonian individualism and identify more general spatial and temporal characteristics of populations that hold within and across species.

GEOGRAPHIC RANGE BOUNDARIES OF MAMMALS

The same picture of emergent general features despite underlying Gleasonian individualistic details is revealed by studies of the geographic range limits of North American land mammals. In one study (Brown and Gibson 1983), I mapped the northernmost limit of the geographic range of each of the families of mammals of South American ancestry that have colonized North America since completion of the inter-American land bridge about 3 million years ago. These mammals do not show correlated range boundaries reflecting their common South American origins (fig. 8.3; for a similar pattern in fishes see Miller 1966). Instead, they exhibit highly individualistic distributions, with some families barely crossing into tropical Central America and others extending as far northward as the cold temperate regions of northern Canada. Note that this individualism at the level of families also holds at the level of species, because the northern range boundary for a family is usually also the edge of the range of the one northernmost species. Thus, this result is reminiscent of the individualistic local and regional population dynamics of breeding birds documented above.

A more general pattern emerges from another study, this time of southern range boundaries, that documents a coordinated pattern of temporal variation in multiple species. Jennifer Frey (1992) has compiled convincing data showing that several species of small mammals have expanded their geographic ranges on the Great Plains southward into Kan-

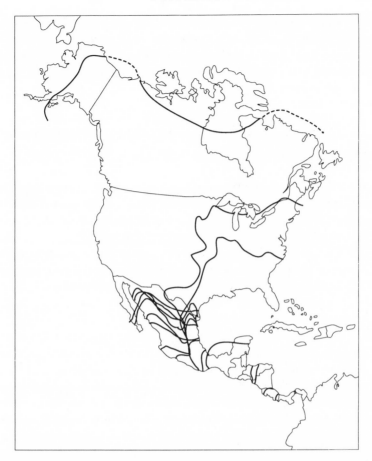

FIGURE 8.3. Approximate northern limits of the geographic ranges of families of terrestrial mammals of South American origin that have dispersed northward into North America since the completion of the inter-American land bridge about 3 million years ago. Note the extremely individualistic responses of these lineages: some families have barely entered tropical Central America, several have spread to varying degrees into subtropical Mexico, and a few (opossums, armadillos, and porcupines) have colonized far into the north temperate region.

sas and Missouri within the last few decades. Three things are particularly interesting about her study. First, although the causes of these shifts are difficult to assign with certainty, they probably reflect some combination of changing climate (e.g., cooler and/or wetter conditions) and human land use practices (e.g., increased irrigation agriculture in this normally dry region). Second, these changes are just the opposite of those that would be expected in response to global warming. Third, these southward shifts represent coordinated changes in the ranges of several species that are widely distributed in cool, mesic habitats over the northern part of the continent. Thus, they are similar to the examples of changes in abundance or distribution of multiple species of birds in response to climate and food on the wintering grounds or to predation on the breeding grounds mentioned above.

POPULATION PROCESSES AND SPECIES DIVERSITY

Gleasonian Individualism, Niche Differentiation, and Coexistence

Our analyses of BBS data suggest that the uniqueness of species niches and the resulting Gleasonian individualism in response to spatial and temporal environmental variation may play a major role in maintaining the diversity of species. The small insectivorous passerines are one of the most species-rich avian guilds, and they make up a large proportion of the species near the mode of the frequency distribution of body sizes (see fig. 5.2b). The patterns of spatial and temporal dynamics that can be induced from the BBS data—increases in some parts of the range offsetting decreases in other parts, and increases in some species in some regions offsetting decreases in others—are just what we would expect if the individualistic differences among species facilitate their coexistence in the continental avifauna.

These varied and often complementary dynamics thus appear to reflect niche differences that enable the insectivorous bird species to subdivide the space and resources of the continent. The pattern of spatial and temporal variation in the environment is such that nearly all species, even those that are relatively rare and geographically restricted, are able to increase in some parts of their ranges. Although we have not identified the specific niche characteristics and environmental conditions that govern these dynamics, they almost certainly do not reflect purely random variation. Random processes would tend to lead rapidly to the extinction of rare, restricted species. If such species are to persist, they must have special requirements that confer some kind of density dependence and enable them to increase when they become rare.

Range Boundaries, Interspecific Interactions, and the Dobzhansky-MacArthur Hypothesis

Robert MacArthur (1972b) proposed the intriguing hypothesis that many North American species are limited at the northern edges of their ranges by physical conditions (see above), but at their southern boundaries by biotic interactions. He further suggested that the southern boundaries usually could not be attributed to exclusion by any single species of competitor or predator, but to the diffuse effects of many species. This hypothesis is appealing because it is consistent with the facts that species diversity increases and the apparent severity of physical conditions decreases toward the wet lowland tropics. Actually MacArthur's hypothesis can be considered a special case of an old idea (Dobzhansky 1950) that biotic factors are more limiting in the tropics, whereas abiotic conditions are more important at higher latitudes—and also, by extension, at higher elevations and in more arid regions, where species diversity is also lower and physical stress appears to be more severe.

This idea has important implications for Rapoport's rule (Rapoport 1982; Stevens 1989, 1992; see also chap. 6) that along the gradient from the tropics to the poles, geographic ranges tend to increase in area and to span a greater range of latitudes. But why should limitation by physical conditions be correlated with wide geographic distributions, whereas limitation by other organisms is associated with more restricted ranges? Consider the following argument. Species that live at high latitudes, high elevations, or in arid regions must be tolerant of a wide range of physical conditions, because even a local population inhabiting a very small area experiences a great deal of temporal variation in the abiotic environment; this is very similar to Janzen's (1967) reason "why mountain passes are higher in the tropics." Such species should be able to have wide geographic ranges, unless they are limited by biotic interactions. Because they occur in regions of low species diversity, however, these species will tend to encounter relatively few other species that might limit their abundance and distribution. On the other hand, species that live in regions of high species diversity encounter many more potential competitors, predators, parasites, and diseases. Some of these other species, or several of them in combination, are likely to overlap in use of resources, interfere with reproduction, or cause mortality and thereby limit abundance and distribution. If the combined effects of biotic interactions with many coexisting species are severe, they could often limit abundance and distribution below levels set by abiotic stresses.

It is probably not this simple. One might ask why a broadly tolerant species does not increase and disperse until it is limited by some biotic interaction, especially since it would seem to be easier to adapt to abiotic stresses, which may be relatively predictable, than to other species, which can coevolve. To some extent this probably does happen. Interspecific interactions can sometimes be important limiting factors in physically variable environments with low species diversity. Thus, some of the best examples of competition and predation limiting the abundance and distribution of species come from experiments in intertidal zones and deserts at temperate latitudes (e.g., Connell 1961a,b, 1975; Paine 1966, 1974;

Lubchenco and Menge 1978; Brown and Munger 1985; Brown and Heske 1990b; Heske, Brown, and Mistry 1994). The effects of other species do not eclipse those of physical conditions, however; many of these investigators have shown at least equally large effects of abiotic factors on the same species.

On the other hand, one could ask whether species in high-diversity, physically more constant environments do not have such narrow tolerances for physical stress that they are also limited by abiotic factors. It may be that this is the case, but the effects of physical variables are just too subtle to be detected. It may also be, however, that even species in relatively constant physical environments tend to have broad tolerances, at least sufficiently broad to occupy much larger ranges than they actually do. That at least some of these species are subject largely to biological control is suggested by the success of some introduced tropical species: in the absence of their natural enemies, some tropical invaders have occupied a much wider range of habitats and attained much higher abundances than in their native regions. An example is a species of *Acacia* that is not a dominant plant in its native Mexico but is converting large expanses of tropical Australia into a shrubby monoculture.

Thus, I have tried to extend what I might call the Dobzhansky-MacArthur hypothesis to suggest how species diversity and physical environmental variation might interact to determine the abundances and distributions of species along environmental and geographic gradients. Note that species diversity is hypothesized to act through biotic interactions to limit abundance and restrict geographic distributions, but of course there is feedback. If the species on average have lower population densities and narrower distributions, then, everything else being equal, the environment should be able to support more species.

How do we escape from this circle, and in particular, why are certain environments, such as the wet tropics, so much richer in species? First, the argument is not totally circular, because outside of the tropics extreme, fluctuating environments and physical limiting factors are hypothesized to play a major role in limiting abundance and distribution, and hence in contributing to low species diversity. Second, I suggest that everything else is not equal, and that factors other than the numerous biotic interactions contribute to the high species diversity in the tropics. In particular, the high productivity of the wet tropics relative to many other environments probably plays a major role in supporting the high diversity of species. Of course, productivity ultimately depends on features of the abiotic environment, of which some of the most important are the intensity and seasonal variability of solar radiation and the quantity and seasonality of precipitation.

One problem with the Dobzhansky-MacArthur hypothesis is that it is difficult to test definitively. It may be possible to identify single physical factors that limit distributions outside the tropics (Root's work on birds is an example), but it will be much more difficult to identify the multiple species and their interactions that might limit populations in the tropics.

Little progress will be made, however, if the difficult challenges are avoided. One thing in favor of the Dobzhansky-MacArthur hypothesis is that it offers a single, parsimonious explanation for general, taxonomically widespread phenomena, such as the low-latitude edges of ranges and both the latitudinal and elevational versions of Rapoport's rule (Stevens 1989, 1992). The hypothesis also seemingly explains the tendency in intertidal communities for abiotic conditions to set upper limits on distributions while biotic interactions tend to limit abundance and distribution lower in the intertidal zone, where species diversity is also higher (e.g., Connell 1975). The susceptibility of islands to colonization by exotic species also emerges as a special case of the Dobzhansky-MacArthur phenomenon. The low species diversity and the lack of many kinds of competitors, predators, and pathogens on islands almost certainly results in lowered biotic resistance to invasions (e.g., Williamson 1981; Moulton and Pimm 1986; Hengeveld 1989; see also chap. 12 and fig. 12.1).

The Dobzhansky-MacArthur hypothesis also seems to be consistent with many anecdotal observations, such as the role of certain diseases in excluding particular species from large areas in the tropics: for example, effects of avian malaria on native birds at low elevations in Hawaii (Van Riper et al. 1986) and exclusion of cattle by the tsetse fly and sleeping sickness from parts of tropical Africa (e.g., Mulligan 1970). It may be possible to obtain other evidence to evaluate the hypothesis. For example, there should be data to test the prediction that more kinds and higher incidences of diseases and parasites will be found in humans and domestic animals and plants in the lowland moist tropics than in high-elevation tropical, high-latitude temperate, or arid regions. It would also be interesting to compare the frequency and apparent causes of both successful and failed introductions of exotic species in tropical and temperate habitats.

CONCLUDING REMARKS

In this chapter I have considered the role of population processes in macroecological phenomena. The search for mechanisms at the population level to explain patterns in assemblages of many species at geographic spatial scales has produced two different, and at first glance conflicting, results. On the one hand, there appears to be a great deal of Gleasonian individualism in the spatial and temporal population dynamics of species in some assemblages, such as breeding North American land birds. On the other hand, other examples of correlated population fluctuations over space, over time, and between species suggest that certain environmental limiting factors and population dynamic processes have general effects.

In fact, as I indicated in chapters 2 and 3, these two characteristics of population phenomena are not contradictory, but are examples of common features of complex systems. The structure and dynamics of the components of these systems (different populations of the same or different species) exhibit idiosyncratic differences in their details—hence the Gleasonian individualism. At the same time, there are emergent characteristics exhibited by the larger systems that are collections of these components (species comprising many local populations, or biotas comprising many species populations)—hence the general patterns. Like the kinds of MacArthurian community structure discussed in chapter 3, however, the characteristics of population dynamics described above do not appear to have the striking degree of regularity and generality as some emergent properties considered in chapters 4, 5, and 6.

One possible exception, which may have both broad generality and important macroecological implications, is the Dobzhansky-MacArthur phenomenon: the tendency, as species diversity increases, for biotic interactions to become increasingly important in determining the abundance, distribution, and ultimately the diversity of species. This hypothesis potentially offers a single parsimonious explanation for such different empirical patterns as Rapoport's rule, the low-latitude and lower-elevation limits of species ranges, the factors limiting abundance and distribution of species in intertidal habitats, and the differential invasibility of island and continental biotas by exotic species.

9 Mechanisms

Species Dynamics

MACROSCOPIC OR TOP-DOWN PROCESSES

In the previous two chapters I discussed how processes at the individual and population levels affect large-scale patterns of abundance, distribution, and diversity. It is neither profitable nor necessary, however, to try to explain all macroscopic phenomena in terms of the mechanisms at lower levels.

Consider the classic question of community ecology: What determines the abundance, distribution, and identity of species within some local habitat? Several investigators have pointed out that a large part of the answer is concerned with the processes that limit membership in the local community to some subset of the species potentially able to colonize from the regional species pool. This problem can be addressed largely by reductionist research that considers the relationship between the local abiotic environment and the requirements and interactions of the species in the pool. But such an approach cannot answer one fundamental question: What determines the composition of the regional species pool? Addressing this question requires entry into the realm of biogeography and macroevolution. It requires an investigation of species dynamics at large spatial and long temporal scales.

CONCEPTS, PROCESSES, AND MODELS

The Analogy with Population Dynamics

I use the term "species dynamics" to refer to changes in the diversity of species in a biota and to the processes of speciation, colonization, and extinction that underlie these changes. Thus, species dynamics are in many ways analogous to population dynamics, which refers to changes in

155

the abundances of individuals in a population, and to the birth, death, and dispersal processes that underlie these dynamics.

The utility of any analogy is limited. For those who are not used to thinking about the regulation of diversity but are familiar with the concepts, processes, and models of population regulation, however, some valuable insights can be gained from their similarities. First, both birth and speciation are basically multiplicative processes. In each case, a single parental unit has the intrinsic capacity to produce more than one offspring unit. In the eighteenth century, Malthus pointed out that this capacity confers on all populations the inherent potential to increase exponentially. This potential will be realized except when environmental conditions limit the survival, dispersal, and reproduction of individuals. Similarly, the number of species in any biota has the inherent capacity to increase exponentially, and diversity will increase in this manner unless environmental conditions limit speciation and colonization and cause extinction.

Second, neither individuals nor species continue to proliferate exponentially for long. In fact, over long periods of time the rate of increase must average very nearly zero. Consequently, there are similarities between the regulation of diversity and population regulation. It will often be appropriate to model both kinds of regulation as the maintenance of a dynamic equilibrium. On the positive side, such equilibrium models tend to use relatively simple mathematics to offer insights into the opposing processes of proliferation and loss. On the negative side, such equilibrium models are often gross oversimplifications of much more complex dynamics in which the numbers of both individuals within populations and species within biotas may fluctuate widely.

Third, just as population dynamics can be considered in terms of the effects of density-independent and density-dependent processes, species dynamics can be said to be affected by diversity-independent and diversity-dependent processes. In the 1960s population ecology was embroiled in a long, ultimately unproductive argument about whether population regulation was fundamentally caused by density-independent or density-dependent processes. This debate assumed a nationalistic fervor, with the Australians Andrewartha and Birch (1954) on one side and several noted American ecologists (e.g., Smith 1961, 1963) on the other. As is usually the case in such arguments, neither side was completely correct; the answer was more complex. We now know that:

1. Fluctuations in abiotic factors, such as weather conditions, which are clearly independent of population density, can cause large changes in the number of individuals.

2. The effect of biotic interactions, both within and among species, usu-

ally varies with population density. Often, but by no means always, these tend to permit increases when populations are low and to cause decreases when they are high.

3. Biotic and abiotic factors can interact to influence population dynamics in a variety of ways. For example, severe weather can cause extensive mortality, but the existence of a limited number of refuges may cause a density-dependent pattern of survival.

4. While there was often a tendency to view density-independent processes as inherently stochastic and density-dependent ones as fundamentally deterministic, this is simply incorrect. Some abiotic factors, such as seasonal weather conditions, are highly predictable, whereas some biotic interactions can be quite unpredictable. Further, recent mathematical explorations of nonlinear dynamics or "chaos" indicate that highly deterministic processes and small differences in initial conditions can cause patterns of fluctuation that appear to be random (e.g., Schaffer and Kot 1986a,b; Hastings et al. 1993).

All of these points are applicable to species dynamics. Presumably, they will be developed as we learn more about the relationship between macroecology and macroevolution. We can also expect to find that the analogy to population dynamics will at some points begin to break down.

The Equilibrium Theory of Insular Biogeography

The first real theory of species dynamics was the equilibrium theory of insular biogeography. Although this theory is usually credited to MacArthur and Wilson (1963, 1967), Munroe had independently developed similar ideas fifteen years earlier (Munroe 1948, 1953; Brown and Lomolino 1989). The theory suggests that the number of species inhabiting an island or isolated patch of habitat represents a dynamic equilibrium between opposing rates of origination and extinction (fig. 9.1a). Rates of colonization are assumed to decrease with increasing isolation or distance from the nearest mainland or other source of colonists, while rates of extinction are assumed to decrease with increasing size and habitat diversity of the island (often measured as area) (fig. 9.1b). Although both Munroe and MacArthur and Wilson recognized that, especially on large islands, new species could come from both immigration from other sources and speciation within the island, their deliberately simplified models attributed origination to colonization.

The simple graphical representations of the regulation of insular species diversity shown in figure 9.1 are the legacy of MacArthur and Wilson (1967). While this model has been criticized, it has played a major role in the revitalization of biogeography that has occurred in the last few decades (see Simberloff 1974, 1980; Gilbert 1980; Brown 1981, 1986, 1988;

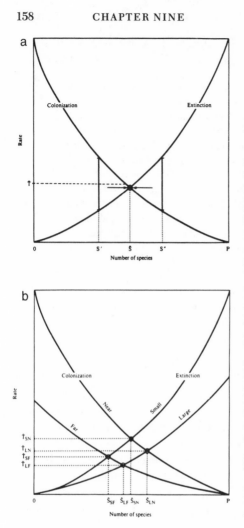

FIGURE 9.1. The MacArthur-Wilson (1967) model for the regulation of species diversity on islands. (a) The simplest model, showing the rate of colonization decreasing and the rate of extinction increasing as a function of the number of species present on the island. The point where the lines cross is a stable equilibrium because if the number of species decreases, the rate of colonization will exceed the rate of extinction, and vice versa if the number of species increases above the equilibrium number. (b) The effects of island size and isolation on the regulation of insular species richness as proposed by MacArthur and Wilson. Island size is hypothesized to affect only the extinction rate, resulting in higher extinction and fewer species as insular area decreases. Isolation is hypothesized to affect only the colonization rate, resulting in decreased colonization and fewer species on more isolated islands. (From J. H. Brown and A. C. Gibson, *Biogeography*, Copyright (c) 1983 Mosby-Year Book, Inc. Reprinted by permission of Wm. C. Brown Communications, Inc. Dubuque, Iowa. All rights reserved.)

Brown and Gibson 1983; Williamson 1981). The MacArthur-Wilson (M-W) model predicts patterns of variation in species richness with respect to island size and isolation. Most research programs in insular biogeography begin by testing these predictions. Because they are robust and qualitative, they are easily falsified. Many advances in insular biogeography have come from developing alternative models to explain cases in which the predictions do not hold (e.g., Brown 1971a; Brown and Kodric-Brown 1977; Simberloff 1980).

Several investigators have tried to generalize the M-W model to noninsular systems; for example, to explain patterns of species richness over latitudinal or elevational gradients on continents or changes in diversity over time (e.g., Rosenzweig 1975, 1992; Brown 1988). One important feature of such treatments is the explicit incorporation of speciation, instead of or in addition to colonization, as an important origination process.

EXTINCTION

Demography

Extinction of a species, like the death of an individual, represents the termination of an important unit of biological organization and a lineage of genetic information. Modern humans have caused enough extinctions that we should have had ample opportunity to study the process. Indeed, much more is known about extinction than about colonization and speciation, the other two processes that contribute to species dynamics.

Ultimately a species goes extinct when its last individual dies, but this is the relatively trivial endpoint of a much more complicated and interesting chain of events. Whenever the death rate exceeds the birth rate, extinction is inevitable unless this trend is reversed. Most extinctions are caused by a combination of demographic population processes and environmental changes. The demographic processes have been well studied because they lend themselves to investigation by mathematical and computer simulation models (e.g., MacArthur and Wilson 1967; Goel and Richter-Dyn 1974; Leigh 1975; Goodman 1986; Pimm 1991). Most of the results are fairly straightforward and intuitive.

First, the probability of extinction increases nonlinearly with decreasing population size. In very small populations chance events, such as otherwise unlikely sex ratios and age structures, can result in a relative or absolute failure to reproduce.

Second, highly fluctuating populations have a higher probability of decreasing to zero than relatively constant populations of comparable average size. While many fluctuations may be caused by environmental change (see below), they can also be influenced by intrinsic demographic characteristics, such as high birth and death rates.

Third, certain density-dependent demographic processes can affect the probability of extinction. Any kind of negative feedback that tends to increase birth rate or decrease death rate as population size decreases will decrease the probability of extinction. Conversely, Allee effects, or positive feedback mechanisms, that tend to reduce birth rates or increase death rates in small populations will increase the probability of extinction. Examples of Allee effects include loss of genetic variability and inbreeding depression, the need of sexual organisms to find appropriate mates, and the role of aggregations of individuals in defense against certain kinds of predators.

The above demographic factors theoretically operate even in a constant environment. The intrinsic characteristics of many populations are such that there is some finite probability of extinction. Since the effects of all but the density-dependent processes are essentially random, demographers and conservation biologists often refer to the resulting population fluctuations as "demographic stochasticity." This term calls attention to the fact that some negative-feedback, density-dependent mechanism is required to prevent a random walk that will eventually end with either extinction or an infinitely large population size.

Environmental Change

The demographic processes described above may contribute importantly to extinctions, but usually they act in conjunction with environmental change. Some kind of deterioration in environmental conditions causes an increase in death rate or a decrease in birth rate and a resulting decrease in population size to the point where the demographic factors come into play. Most populations do have negative feedback mechanisms that tend to cause recovery after numbers have been reduced. But environmental changes can overwhelm these mechanisms and push populations to critically low levels.

The fossil record shows that species have been going extinct throughout the history of life on earth. More than 99.99% of the species that once lived have disappeared. Paleontologists and macroevolutionists distinguish, at least at the extremes, two kinds of extinctions. They refer to the low to moderate rates of disappearance of lineages throughout the fossil record as background extinctions. In contrast, they refer to the high rates of extinction that occurred during relatively short intervals of time, and usually occurred synchronously in many different kinds of organisms, as mass extinctions (e.g., Raup 1979, 1986; Raup and Sepkoski 1982, 1984; Sepkoski 1982, 1986; see also Quinn 1983).

The fossil record provides few clues to the causes of most background extinctions. In fact, the seeming randomness of the disappearance of lineages led Van Valen (1973b) to advance the "Red Queen" hypothesis. Van Valen suggested that evolution is a coevolutionary race in which those species that fail to keep up with the competition and predation from all the other species are continually going extinct. While there has been considerable discussion of the Red Queen hypothesis, it is hard to document coevolution—much less the failure of species to coevolve—in the fossil record.

There is much stronger evidence, however, that several of the mass extinctions were caused by rapid, worldwide environmental change (e.g., see Jablonski 1986a). Stanley (1984) has suggested global climate change, especially episodes of cooling and glaciation, as the cause of several mass extinctions. Perhaps the two largest extinction events in the fossil record occurred at boundaries between the Permian and Triassic and the Cretaceous and Tertiary epochs. The causes of the former event remain obscure and controversial. There is now good evidence that the latter event was triggered by the impact on the earth of a massive asteroid, which injected an enormous cloud of particulates into the atmosphere, caused rapid global cooling, and triggered the near-simultaneous extinctions of dinosaurs, ammonites, and many other lineages (Alvarez et al. 1980, 1984). Perhaps the most convincing evidence is the recent discovery of the crater of the asteroid off the coast of Yucatán (Swisher et al. 1992).

Undoubtedly the best evidence that environmental change is a major cause of extinction, however, is the wave of human-caused extinctions that began in the Pleistocene and is still occurring. There can be no doubt that humans have caused the extinction of a large number of plant and animal species, both directly by predation and habitat alteration and indirectly through the introduction of domestic and exotic species, pollution, and perhaps even global climate change. But discussion of these human impacts will be deferred until chapter 12.

How Predictable Is Extinction?

There has always been an element of contradiction in the way that scientists have viewed extinction. There has usually been a tendency, both in the interpretation of data and in the development of mathematical models, to treat extinction as a stochastic process. The demographic and environmental processes that cause extinctions are viewed as acting unpredictably in space and time, capriciously eliminating some species and allowing others to survive. On the other hand, there has always been a tendency to search for the cause of extinction events. Now there is increasing evidence that, even though extinction is fundamentally a probabilistic process, it is surprisingly predictable. When investigators have studied extinction quantitatively, either by comparing the rates of extinction in different kinds of organisms or by comparing the traits of species that have gone extinct and survived under the same environmental conditions, they have almost invariably found clear patterns.

Perhaps the clearest examples come from nested subset species compositions of islands and insular habitats (see chaps. 1 and 3). There are many archipelagoes where the species inhabiting the individual islands have been derived from a common biota by extinctions that followed isolation at the end of the Pleistocene. Examples include islands on the continental shelves isolated by rising sea levels as the glaciers melted, and the mountaintop islands of southwestern North America isolated by the retreat of cool, mesic forest habitats to higher elevation as the climate warmed. In most of the cases that have been studied (reviewed by B. D. Patterson et al., unpub.), the extinctions have been highly predictable, leaving a nested subset community structure on islands of varying size (e.g., table 1.1). Furthermore, the extinction-proneness of the species varies in predictable ways with traits of the organisms that affect population size. Thus, species of large body size, carnivorous diet, and/or specialized habitat requirements have typically disappeared from all but the largest islands (Brown 1971a).

Large body size repeatedly appears as a characteristic associated with high rates of extinction during episodes of environmental change. The dinosaurs and other clades of giant reptiles went extinct at the Cretaceous-Tertiary boundary. The extinctions in North America during the late Pleistocene differentially eliminated mammals of large body size and open steppe or grassland habitat (Martin and Klein 1984; Owen-Smith 1989). The birds that disappeared at about the same time included a disproportionate number of carnivores and scavengers, as well as species of large size and open habitats (Grayson 1977, 1984). The extinctions and near extinctions caused by modern humans during the last two centuries also differentially affected the largest vertebrates: the great auk, Steller's sea cow, giant tortoises, sea turtles, rhinoceroses, and whales (e.g., Diamond 1984).

Large size has at least two correlates that should make for susceptibility to extinction. First, as mentioned in chapter 5, large size severely constrains population density and hence total population size. Large organisms are less common than their smaller relatives, and thus any environmental changes that further reduce their populations increase their probability of extinction. Second, large organisms have life history traits that make them susceptible to environmental change. Many large birds and mammals produce small clutches or litters of relatively large offspring with long intervals between clutches. Other organisms, such as many trees and fishes, produce very large clutches of tiny offspring, but these suffer enormous mortality during the long time required to grow to maturity. In either case, any environmental change that causes additional mortality or reproductive failure may change the situation to one in which birth rates are no longer barely adequate to offset death rates, so that extinction becomes inevitable. It should be noted, however, that these same life history traits of large organisms will tend to reduce the magnitude of fluctuations in population size. At least under relatively stable environmental conditions, such constancy may reduce the probability of extinction.

Two recent analyses of time series of populations inhabiting very small islands have been able to relate the incidence of extinction to demographic characteristics of populations. Pimm, Jones, and Diamond (1988; see also Tracy and George 1992), analyzing censuses of breeding birds on small islands off Britain, and Schoener and Spiller (1987, 1992), monitoring spiders on small cays in the Bahamas, both found a higher incidence of extinction in populations that on average were smaller or had wider fluctuations. Not all of their results agree, however, and Schoener and Spiller (1992) present a very cogent discussion of how the apparent patterns of extinction depend on the methods used to analyze the data.

One may question whether the inferences from these studies of tiny islands for at most a few decades can be extrapolated to larger spatial and temporal scales. This is a problem especially for the bird studies, because most of the "extinctions" were recorded in bird species and on islands where the "population" never numbered more than one pair, or at most a few pairs. Furthermore, many of the species are migratory, and on their annual migrations to and from their wintering grounds in Africa they cross much larger stretches of water than those separating their breeding islets from Great Britain. Thus, if one performed annual censuses of birds in small woodlots in England or the central United States, or of spiders in gardens or garages anywhere in the world, one might find that certain species tended to disappear and reappear. The highest incidence of disappearances might well occur in those species and/or census sites where the populations were small and variable. What does this tell us about extinctions of entire species?

I, quite frankly, am not sure. It may be possible to gain insights about patterns and processes at geographic spatial scales and geologic time scales from studies done at much smaller scales. This would be very useful because it is hard to get census data for more than a few decades or to get good data from large regions. Extrapolation beyond the range of parameter values in the data is always hard to justify statistically, however. It is especially hazardous to extrapolate over the several orders of magnitude that would be required to go from tiny islands to entire continents and from decades to millennia.

The analyses of Pimm et al. (1988; see also Pimm 1991) suggest that the probability of disappearance is lower for large birds than for small ones on the tiny islands off Britain. This does not fit well with the apparent susceptibility of species of large size to extinction noted above. This discrepancy might be explained if the observed insular extinctions were caused primarily by demographic mechanisms during periods of relatively constant conditions rather than by major environmental changes. As suggested above, large-sized species may tend to be susceptible to environmental change but to maintain relatively constant populations in stable environments. If this is correct, however, it suggests that it is hazardous to make sweeping predictions about species extinctions and conservation policy from studies of small populations living in relatively stable insular habitats.

Another question about extrapolating from data for small islands is whether the disappearance of local populations can really be equated with the extinction of an entire species. Can metapopulation dynamics, the local colonization and extinction phenomena that normally occur in species that inhabit patchy and especially ephemeral habitats, be extrapolated to species dynamics? In table 9.1 are estimates that I have obtained from colleagues and from the literature of the total populations of rare species that apparently have not been greatly reduced by human activities. It is possible to find additional examples of bird, fish, and plant species, mostly from islands or insular habitats, in which the total number of individuals is in the hundreds. But these represent the smallest known populations for entire species, and they are orders of magnitude larger than the local populations of birds and spiders that went extinct on small

TABLE 9.1. Estimated total population sizes of some relatively rare, restricted plant and animal populations that have not been much reduced by the activities of European humans.

Species	Total population	Geographic range	Source
Hess's fleabane (*Erigeron hessi*)	500–1,000	<1 km^2	E. DeBruin, unpub.
Nodding rockdaisy (*Perityle cernua*)	5,000–10,000	21 km^2	E. DeBruin, unpub.
Organ Mountain evening primrose (*Oenothera organensis*)	3,000–5,000	130 km^2	E. DeBruin, unpub.
Organ Mountain foxtail cactus (*Coryphantha organensis*)	5,000–10,000	130 km^2	E. DeBruin, unpub.
Sangre de Cristo pea-clam (*Pisidium sanguinichristi*)	607,000	263 m^2	P. Mehlhop, unpub.
Chloride oreohelix snail (*Oreohelix pilsbryi*)	500–1,000	12 ha	P. Mehlhop, unpub.
New Mexico hot spring snail (*Pyrgulopsis thermalis*)	3,000–5,000	<1 km^2	P. Mehlhop, unpub.
Devil's Hole pupfish (*Cyprinodon diabolis*)	200–600	200 m^2	Soltz and Naiman 1978

islands. Most continental and marine species appear to have comprised at least thousands of (and often orders of magnitude more than that) individuals before the impacts of modern humans.

These observations suggest that large, infrequent environmental changes are usually required to reduce species populations to levels at which there is a measurable risk of extinction. Sometimes such changes occur regionally (e.g., a volcanic eruption or a severe drought) or in ways that affect only a few species, and these cause background extinctions. Sometimes they occur on a more global scale, affect many species, and cause mass extinctions. As others have pointed out (e.g., Diamond 1984; Wilson 1985, Simberloff 1986), modern humans are causing a wave of mass extinction comparable to those in the fossil record.

SPECIATION

Models

Speciation is the only process that opposes extinction and creates new species. Unfortunately, speciation is also one of the least well understood processes in all of biology. There are many models of speciation. They have been produced by systematists (Mayr 1942, 1963; Brown 1957; Bush 1969, 1975; Brooks and McLennan 1991), geneticists (White 1968; Carson 1968; Carson and Kaneshiro 1976; Templeton 1980), and ecologists (Rosenzweig 1978). Most of them are incomplete. They fail to recognize that the formation of two species from a single ancestral one must be the result of an interaction between the ecological and biogeographic setting and the evolutionary processes that cause genetic differentiation. Perhaps it is not surprising, therefore, that in the great majority of cases of closely related "sister" species, it cannot be determined whether they were formed by allopatric or sympatric speciation—and, if allopatric speciation is suspected, whether it occurred by the vicariant separation of large populations, the splitting off of small, peripherally isolated populations, or the founding of new populations by dispersal across a geographic barrier.

I do not mean to be unduly negative here. Unlike extinction, speciation is difficult to study because human-caused speciation is not going on all around us on a time scale such that many examples of the entire process can be observed directly (but see Bush 1969; Rice and Salt 1990). In order to understand the process of speciation, it is necessary to make inferences from the genetic, ecological, and biogeographic characteristics of populations that appear to represent various stages in the process, from those just beginning to diverge to well-differentiated sister species. Increasingly, investigators are doing just this (e.g., Carson and Kaneshiro 1976; Wiley and Mayden 1985; Brooks and McLennan 1991), but there is still a long way to go.

Ecological and Biogeographic Considerations

I am afraid that I do not have much to contribute beyond a bit of macroecological perspective. It seems clear to me that most speciation is not driven by any intrinsic genetic process that operates independent of the environment. Just as environmental change plays a major role in extinction, so does environmental variation in space and time play a major role in the formation of new species. In order for a lineage to split and for at least two descendant species to survive, it is necessary that ecological and/ or biogeographic, as well as genetic, differentiation occur. If the new species are sympatric, they must differ sufficiently in their ecological requirements so as to coexist. If they are allopatric, each must have ecological niche requirements and a sufficiently large geographic range so as to maintain a viable population and avoid extinction. Unless these conditions are met, the process of genetic differentiation will usually not go to completion, because the incipient species will either fuse or go extinct.

This suggests at least two profitable areas for research. First, comparative studies of closely related species can potentially tell us much about the kinds of differences in morphology, physiology, behavior, ecology, and biogeography that characterize the successful products of speciation. Such studies can be done in two ways. One is to do statistical, macroecological studies of lineages that are much smaller, and hence contain species that are much more closely related, than the birds and mammals that I have studied. While this approach may sacrifice some ability to detect patterns in large data sets, it has the advantage of being better able to interpret the ecological and evolutionary significance of the traits being analyzed. Promising examples of such studies are those of Glazier (1980) on mice of the genus *Peromyscus,* of Martin (1979, 1984) on fossil and recent cotton rats of the genus *Sigmodon,* of Heaney (1984a) on squirrels of the family Sciuridae, of Gittelman (1985) on the mammalian order Carnivora, and of Schluter (1984) on Darwin's finches. Alternatively, comparative studies can be done in a strict phylogenetic framework, so that the most closely related species are identified correctly. The recent studies of Brooks and McLennan (1991), Harvey and Pagel (1991), Cotgreave and Harvey (1992), Taylor and Gotelli (in press), and others represent promising beginnings.

One note of caution is in order. In such comparative studies there is often the temptation to assume that the present species reflect all of the important speciation events that occurred in a clade, and hence that the different characteristics of those species evolved during or as a direct result of those speciation events. Neither of these assumptions may be correct. The high incidence of extinction occurring today and observed in the fossil record suggests that there were often many more speciation events than there are extant species in a clade. If this is the case, even differences between species that were directly associated with speciation events may have arisen during events that gave rise to species that are now extinct (fig. 9.2). Furthermore, even if all products of a speciation event have survived, some of the differences among them may have originated long after the speciation and may be at best only indirectly related to the speciation process.

A second and related promising line of investigation is the comparative study of the radiation of different lineages in the same environment or of the same clade in different environments. Some of the most spectacular examples of cladogenesis are from isolated islands or archipelagoes. As interesting as the high incidence of speciation in these lineages, however, is the lack of speciation in other lineages in the same ecological and biogeographic setting. Why have finches radiated in the Galápagos while mockingbirds have not? Why have just a few lineages of cichlid fishes repeatedly speciated to give rise to enormous species flocks in the Great

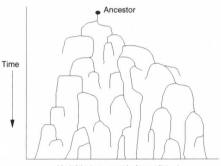

Time

Variable (e.g. morphology, diet ...)

Ancestor

FIGURE 9.2. A hypothetical diagram showing the proliferation of a clade from a common ancestor and differentiation of the species with respect to some quantitative trait, such as body size. The pattern that I have drawn is not strictly "punctuational" and depicts considerable anagenetic change between speciation events. The point here, however, is that in the absence of an excellent fossil record, these anagenetic changes, and especially extinction events, make it difficult to infer changes in character states that were associated with particular speciation events. Cladistic techniques applied to the contemporary species may permit reconstruction of the relative temporal sequence of lineage branching (speciation events) that gave rise to those species, but they will miss speciation events—and changes in character state associated with those speciation events—that were followed by extinction of one or more of the descendant species.

Lakes of central Africa, while other fish groups that have apparently been isolated in the same lakes for the same length of time have persisted but have not multiplied? Why has *Drosophila* radiated so much in the Hawaiian Islands, but not in the Lesser Antilles or other tropical archipelagoes that it has colonized? These questions focus attention squarely on the role of interactions between species and their environments. I suspect that many of the answers lie in the relationship between organism-specific traits, such as mobility and niche requirements, and environment-specific features, such as degree of habitat heterogeneity and the history of environmental changes. I will consider these factors briefly at the end of this chapter when I discuss continental patterns of species diversity.

COLONIZATION AND RANGE SHIFTS

Processes

Speciation and extinction are the only two processes that can alter global species diversity. But movement of individuals into new areas or out of previously inhabited ones can also have important effects on diversity on local to continental scales. Dramatic examples occur when new habitats are created and colonized, such as after the volcanic eruptions on Mount Saint Helens and the island of Krakatoa, or when long-standing barriers are removed, such as after completion of the land bridge connecting North and South America in the Pliocene.

At the extremes, there are two kinds of movements: immigration and emigration at the scale of local habitats, and larger-scale expansion and contraction of the geographic range. Both kinds can occur either by movement of individuals into essentially contiguous areas or by dispersal across uninhabitable areas or geographic barriers.

Successful colonization actually involves two processes: dispersal and establishment. These may be difficult to distinguish, especially when individuals move into essentially contiguous areas. For example, one could question whether sink populations, sustained by immigration because death rates exceed birth rates, should really be considered established, even though they may appear to be relatively permanent (Pulliam 1988). When dispersal is a relatively infrequent event that involves crossing barriers of unsuitable environment, however, then immigration has no lasting effect unless the dispersing individual not only survives but leaves descendants. In such cases, the minimum colonizing unit is what biogeographers call a propagule. Except in asexual organisms, which tend to be excellent colonizers because a single individual can found a new population, the critical establishment phase requires the coincident arrival and survival of multiple individuals of appropriate ages and sexes. This need for a propagule places additional constraints on the traits of organisms that are able to cross barriers of different kinds.

How Predictable Is Colonization?

Dispersal, like extinction, has traditionally been considered a highly stochastic process. Increasingly, the evidence suggests that dispersal is probabilistic but quite predictable. The best evidence that I know of comes from studies by Mark Lomolino (e.g.; 1986, 1989, 1994a,b) on the nonvolant mammals inhabiting small islands in the Saint Lawrence River and in Lake Huron. Lomolino has developed a multivariate statistical technique to analyze the presences or absences of individual species as a function of island isolation and area. This enables him to assess the contributions of colonizing ability and extinction resistance to the pattern of distribution. As a result he has identified species that are good and poor immigrators and has shown that they possess combinations of traits that either facilitate or inhibit dispersal by swimming and rafting in summer or by crossing ice in winter. He is also able to identify species that are either good or poor survivors, and to find traits that are associated with this variation in extinction resistance.

Perhaps the most direct evidence of shifts in geographic range boundaries comes from recent changes that have occurred as a result of human activities. Some of the advances and retreats are extensive. Several native bird and mammal species, including the cardinal, mockingbird, tufted titmouse, and opossum, have expanded their geographic ranges in eastern North America tens and even hundreds of kilometers to the north in the last several decades (e.g., Boyd and Nunneley 1964; Hill and Hagan 1991). While their exact causes may be species-specific and debatable, these range shifts appear to have occurred primarily in response to human-caused changes in habitat and food supply. In southwestern North America, by contrast, several species, including the masked bobwhite, thick-billed parrot, Mexican wolf, and jaguar, have contracted their ranges southward so that they no longer occur north of the U.S.-Mexico Border. Again, the exact causes of these retreats have not been identified, but some combination of human hunting and habitat alteration is undoubtedly primarily responsible.

The literature of exotics is filled with cases in which human assistance in crossing biogeographic barriers has enabled species to colonize distant regions (e.g., Drake et al. 1989; Hengeveld 1989). These clearly represent situations in which only limited dispersal was previously preventing the species from occurring in geographic regions and ecological conditions in which they could otherwise exist.

As in any case of range expansion, long-distance, barrier-crossing dispersal will be successful only if the propagules are able to become established at the new site. Here, we can learn a great deal from studies of the successes and failures of introduced exotic species (Drake et al. 1989; Hengeveld 1989). I will make three points that probably apply to cases of both human transport and natural long-distance dispersal. First, success depends upon a substantial element of apparent chance, especially when the number of colonists is small. This is best illustrated by examples of intentional introductions, in which several attempts may be necessary before one is finally successful. Second, there is a large element of species-specific variability, because the local environmental conditions must meet the niche requirements of the species. This makes it difficult to predict whether a given species will become established at a particular site without a detailed study of its ecology and of the environmental characteristics of the site of colonization. Finally, superimposed on this unpredictability and species specificity are some robust probabilistic rules. These do enable us to predict both the relative colonizing success of different kinds of species and the relative susceptibility of different kinds of environments to invasion. Because the predictability of establishment of exotic species is of considerable importance in conservation, I will defer further discussion of this topic until chapter 12.

THE REGULATION OF SPECIES DIVERSITY

Isolation, Endemism, and Diversity

Perhaps the most important lesson to be learned from the human-assisted spread of exotic species is the extent to which global species diversity is dependent on the fragmentation and isolation of habitats. The earth's land and water are distributed heterogeneously in patches of varying size, isolation, and discreteness: oceans, continents, islands, lakes, mountaintops, streams, marshes, caves, and so on. This fragmentation has promoted speciation and the buildup of endemic biotas.

Isolation is only part of the story, however. Small, isolated islands and other insular habitats have fewer species than nonisolated habitats of the same area. This is readily apparent from a plot of number of species as a function of area for both insular and nonisolated habitats (fig. 9.3). The reason, as MacArthur and Wilson (1963, 1967) pointed out, concerns the balance between immigration and extinction. The higher rates of dispersal among less isolated habitats "rescue" populations from extinction or permit them to recolonize rapidly following any local extinctions that do occur (this is the subject of metapopulation dynamics: e.g., Levins

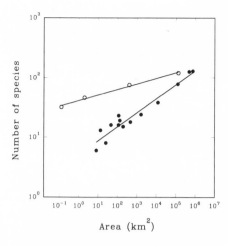

FIGURE 9.3. Species-area relationships, plotted on logarithmic axes, for nonvolant land mammals in isolated and nonisolated habitats. Solid circles, islands of the Sunda Shelf that were connected to the mainland of southeastern Asia until isolated by rising sea levels at the end of the Pleistocene; open circles, nonisolated areas on the southeastern Asian mainland. In the isolated habitats, where there is a much reduced rate of immigration of all species, extinctions occur because of the absence of a "rescue effect"; as a consequence, the islands support fewer species than nonisolated areas of similar size. (Data from Heaney 1984b, fig. no. 1 on p. 12. (c) Springer-Verlag.)

1969; Brown and Kodric-Brown 1977; Pulliam 1988; Gilpin and Hanski 1991). Not sustained by such high rates of dispersal, more isolated populations go extinct, and consequently an insular habitat supports fewer species.

Thus, the overall pattern of species diversity is determined largely by the extent of fragmentation (the number, size, and degree of isolation of patches and the kinds of habitat that they contain), the rates of dispersal, and the extent to which species are endemic to single fragments or specific to particular types of habitat (which depends on the history of and the role of ecological factors in speciation).

Consider two extreme cases. In the first, the habitat is highly fragmented, the isolation has a long history, and dispersal among the patches is very infrequent. This scenario favors low diversity but a high degree of endemism in each patch. This extreme is approached by freshwater fishes. Although all the lakes, rivers, streams, and springs together make up only a tiny fraction of the earth's water, more than half of the world's fish species live in these freshwater habitats. The reason is that isolation has preserved the products of the speciation events that occurred either within single bodies of water or when waters became fragmented and rejoined. Thus, while any given area within a lake or river has fewer fish species than a comparable area of coral reef, each isolated lake or river system tends to have its own distinctive fish fauna, with a substantial number of endemic species. Tiny desert springs, with only a few hundred square meters of water surface, support endemic species.

At the other extreme, the environment is much more continuous, so that different kinds of habitat tend to grade into one another and do not

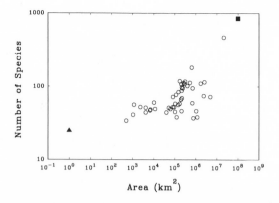

FIGURE 9.4. The species-area relationship, plotted on logarithmic axes, for North American terrestrial mammals in nonisolated sample sites covering a very large range of spatial scales. The solid square represents the entire continent, the open circles represent major biomes (large regions of relatively similar climate and vegetation), and the solid triangle represents the average value for 23 small patches of relatively homogeneous habitat. Note the conspicuously curvilinear pattern. (Data from Brown and Nicoletto 1991.)

have a long history of isolation, and dispersal among them is high. This scenario promotes high species diversity but little endemism within local habitats, and relatively little change in species composition across the landscape. This situation is approached by North American land birds. The lists of sightings for small sanctuaries eventually come to include the majority of species whose geographic ranges encompass the reserve and even some "accidentals" that have dispersed far beyond their normal ranges (see Grinell 1922). The most restricted bird species, such as Kirtland's and golden-cheeked warblers, historically had geographic ranges more than 100 million times larger than those of the desert spring fishes! Most organisms, including the other classes of vertebrates, exhibit patterns of endemism and diversity intermediate between the fish and bird examples.

A Pattern of Continental Diversity

One consequence of limited dispersal, restricted environmental tolerances, and historical speciation is a distinctive shape of log-log plots of species-area relationships for continents (fig. 9.4): an accelerating curve, with the slope becoming steeper as the spatial scale increases.[1] To see why this is so, imagine two extremes. In most cases, increasing the area

1. In compiling such species-area curves, it is important to insure thorough sampling—that is, sampling of sufficient intensity and duration to record all species that occur in the area. Failure to do so will typically underestimate the diversity in the smaller patches (compare my figure 9.4 with figure 38 in C. B. Williams's (1964) book. If sampling were complete, the slope of the species-area curve (when plotted on logarithmic axes) would theoretically

from 1 ha to 10 ha will add very few new species. At this scale, at least for vertebrates, habitats tend to be similar and easily reached by dispersing individuals. On the other hand, increasing the area from one-tenth of the continent to the entire North American continent will result in the addition of many new species. At this scale, many new habitat types are added, dispersal is limited, and there is much endemism (i.e., the geographic ranges of even most mammal and bird species encompass only a small fraction of the area of the continent; see fig. 6.1). It is not clear, however, what combinations of dispersal, habitat specificity, and endemism will produce the highest total continental species diversity. At least in North America, there are more species of birds and freshwater fish at the extremes, than of mammals, reptiles, and amphibians with intermediate combinations.

CONCLUDING REMARKS

Global species diversity is ultimately regulated by the balance between speciation and extinction. At slightly smaller scales diversity is also affected by the spatial distribution of geographic ranges and by the shifting boundaries of these ranges. These processes of species dynamics are fundamentally macroscopic. Speciation, in particular, usually appears to occur on scales of at least thousands of years and thousands of square kilometers. Shifts of geographic ranges and extinctions of entire species can occur rapidly, but when they do, they are usually caused by large-scale environmental change.

All of the processes of species dynamics are probabilistic, but they are also predictable. What ultimately regulates species diversity at all scales is the ways that the dispersal, survival, and proliferation of lineages of organisms, each with its own unique evolutionary constraints and ecological requirements, are influenced by the spatial and temporal variation of the global environment.

---

vary from zero at infinitesimally small scales, because all species are shared, to unity at the largest scales, where there are no species in common.

10 Synthesis

Ecological Implications

HOW GENERAL ARE MACROECOLOGICAL PATTERNS?

How general are the patterns and processes discussed in this book? On the one hand, my ideas and examples are heavily biased by the experience and intuition derived from my own research on mammals and birds. When it is possible to obtain comparable data, many macroecological phenomena appear to be very similar in these two groups of endothermic vertebrates. Is this because the patterns and processes are general, or because birds and mammals are closely related, structurally and functionally similar, and subject to many of the same constraints?

On the other hand, the mechanisms that underlie many of the macroecological patterns—to the extent that we understand them or that I have presented hypotheses to account for them—are potentially very general. They are not specific to terrestrial vertebrates that are endothermic, have high levels of parental care, and possess other traits that are shared by most birds and mammals but not by other kinds of organisms. The physiological and ecological constraints of body size and the fundamental features of population and species dynamics are common to all organisms. Consequently, I suspect that many of the macroecological patterns and processes, with only minor or quantitative modifications, will be equally general.

One problem is that, while comparable data are potentially available for other groups of organisms, few of them have been compiled, analyzed, and interpreted to address macroecological questions. To the limited extent that this has been done, however, the results are encouraging. Two patterns appear to be essentially universal. One is the frequency distribution of body sizes among species shown in figure 5.2. Qualitatively similar relationships have been found in mammals, birds, reptiles, fishes, several groups of insects, trees, and two groups of bacteria (Hutchinson and Mac-Arthur 1959; May 1978, 1986; Bonner 1988). The other nearly universal pattern is the latitudinal gradient of species diversity (e.g., fig. 10.1). Qualitatively similar relationships have been found in mammals, birds, reptiles, amphibians, fishes, grasshoppers, and several groups of vascular plants and marine invertebrates (Fischer 1960; Simpson 1964; Cook 1969; Kiester 1971; Horn and Allen 1978; Otte 1981). Stehli (1968; see also Stehli, Douglas, and Newell 1969; Stehli and Wells 1971) not only demonstrated the latitudinal diversity gradient in marine foraminifera and corals, but also showed that the same patterns have existed for hundreds of millions of years. Related geographic patterns, such as decreases in species richness with increasing elevation and aridity, and the relationship between species diversity and breadth of geographic range called Rapoport's rule have also been demonstrated in a wide variety of organisms (Stevens 1989, 1992). Our own analyses of spatial variation in abundance have revealed qualitatively indistinguishable patterns in several bird, crustacean, parasite, and plant species (fig. 4.5; J. H. Brown, D. W. Mehlman, and G. C. Stevens, unpub.).

Lawton, Gaston, Morse, and colleagues (e.g., Morse, Stock, and Lawton 1988; Gaston and Lawton 1988a,b; Lawton 1990; Basset and Kitching 1991) have demonstrated relationships among body size, abundance, and distribution in insects that are remarkably similar to those in birds and mammals (for example, compare figure 10.2 with figures 5.2 and 5.7a.) This similarity is encouraging because insects differ in so many ways from birds and mammals. Insects are ectothermic and orders of magnitude smaller in body size. Many of them have complex life cycles and use the environment on much smaller spatial and temporal scales than birds and mammals. If insects and endothermic vertebrates are so similar, there is reason to be optimistic that other kinds of organisms will also conform to these patterns.

There is ample opportunity to find out. As mentioned above, the relevant data are available for many groups. While it will be a major task to compile and analyze these data, I hope this book will encourage other investigators to do so. The empirical patterns and the mechanisms proposed for North American terrestrial birds and mammals can serve as hypotheses to be tested for their application to other kinds of organisms and in different habitats and geographic regions.

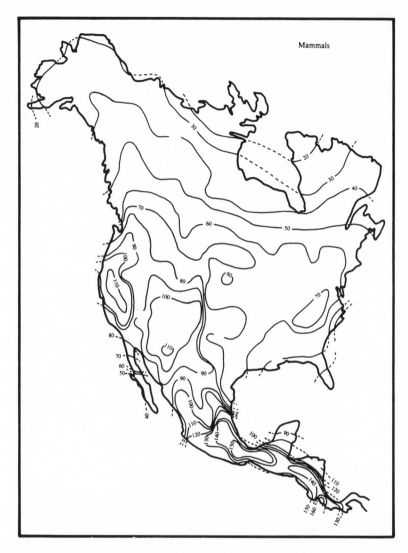

FIGURE 10.1. The geographic pattern of species richness in North American land mammals, showing the pronounced latitudinal increase in the number of species from the poles toward the equator. These data, replotted from Simpson (1984), are based on the number of species inhabiting quadrats 150 miles (approximately 250 km) on a side. (From J. H. Brown and A. C. Gibson, *Biogeography,* Copyright (c) 1983 Mosby-Year Book, Inc. Reprinted by permission of Wm. C. Brown Communications, Inc. Dubuque, Iowa. All rights reserved.)

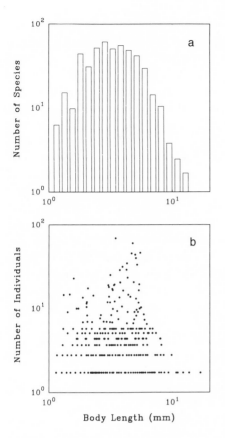

FIGURE 10.2. Two macroecological patterns related to body size in beetles from tropical rain forest in Borneo. *(a)* Number of species as a function of body length, both on logarithmic axes. *(b)* Abundance as a function of body length, both on logarithmic axes. Note the similarity between these patterns and those for vertebrates shown in figures 5.2 and 5.7. (Adapted from Morse, Stock, and Lawton 1988, figs. 4*(a)* and 5*(b)*.)

IMPLICATIONS FOR EXPERIMENTAL ECOLOGY

The Limitations of Ecological Experiments

During the last few decades ecology has grown from its roots in descriptive natural history and developed into a rigorous, sophisticated modern science. I think that two things are primarily responsible for this change. One is the development of ecological theory, especially rigorous mathematical models. It was easy for the natural historian to become enamored of the seemingly infinite variety and overwhelming complexity of living things. The production of theories and models not only gave the discipline a set of logical ideas about how the natural world is organized, it also stimulated empirical scientists to devise rigorous ways to test those ideas.

The second important development was the adoption of modern experimental methods for testing models and hypotheses. Following the success of Connell's (1961a,b) classic barnacle studies, ecologists were quick to embrace the approach of experimental manipulation, the concepts of experimental design, and the tools of statistical analysis. This experimental approach has served the discipline well for three decades. There is no denying that it has resulted in important advances in our understanding of the abundance, distribution, and diversity of organisms and of the fluxes of energy and materials through ecosystems.

In fact, this experimental approach has seemingly been so powerful that many ecologists now seem to think that controlled, replicated manipulations are the only way to do sound empirical research. They are unaware of, or have ignored, the fact that the practical constraints of experimental manipulation are causing them to work at small spatial and temporal scales, thereby giving them a severely limited view of ecological phenomena. For example, a review of experimental field studies in population ecology by Kareiva and Andersen (1988) found that the average size of experimental plots was less than one square meter. A similar review by Tilman (1989) reported that 40% of field experiments lasted less than one year. Modern experimental ecology has become the study of small places for short periods.

Things are changing. Throughout the last few decades some ecologists continued to do excellent nonexperimental research, especially at large scales at which manipulation is impractical. The big change, however, has been fueled by the necessity to confront practical environmental problems, such as changes in atmospheric gases and global climate, depletion and pollution of ground water, destruction and fragmentation of habitats, and loss of biological diversity. To address these problems at the regional to global scales at which they are most severe, ecologists have increasingly needed to develop nonmanipulative approaches.

In addition to its limited scale, experimental field ecology has three other serious problems. The first is the creation of possible artifacts. The human intervention necessary to control and manipulate the system and gather the data may alter the system, and may even cause the results attributed to natural processes. The second problem is Type II statistical error. Because of shortages of resources, the replication and statistical power of ecological experiments are nearly always severely limited. Thus, the experimental design, data, and analyses may not be sufficient to detect small but important responses to the manipulations (for example, see Dueser, Porter, and Dooley's 1989 analysis of experimental studies of competition). The third problem is the difficulty of assessing the generality of the results. If the experiments have been done rigorously and without artifacts, the results may confidently be inferred to have occurred in that place at that time. But will the same results be obtained if the experiment is repeated next year, in a slightly different habitat, or a few hundred kilometers away? Obviously, this is an important question. Equally obviously, the only way to answer it is to replicate the experiment in space and time. But ecological experiments are rarely repeated, and when they are, often the same results are not obtained in other environmental contexts (e.g., Menge et al., in press).

Lest these seem to be unimportant concerns, let me show that they are not by giving examples from my own experimental research on interactions among desert granivores and plants (see Brown et al. 1986 for a review). First, the fences with different-sized holes used to enclose the plots and to exclude particular combinations of rodent species may create an important artifact: they also probably exclude large carnivorous snakes, potentially confounding the effects of competition and predation. Second, spatial heterogeneity, practical constraints on sampling, and limited replication conspire to create many Type II error problems. For example, we are able to document dramatic changes in plant species composition in response to rodent removal, but only years after these changes have begun and have been detectable by eye. Third, how far the results can be generalized will always be a question. When we performed similar experiments, reciprocally removing rodents and ants, in two desert habitats separated by about 200 km, we obtained very different results: rodents and ants appeared to compete strongly at the Sonoran but not at the Chihuahuan Desert site (compare Brown and Davidson 1977 with Brown et al. 1986; Valone, Brown, and Heske 1993). So which result is correct, and if, as we suspect, both results are valid, then what causes the variation in the strength of the interactions?

Problems of Scale and Openness

I am not trying to imply either that there are fatal flaws in the small-scale, experimental, reductionist approach to ecology or that adopting a macroecological approach can solve all these problems. I do suggest that macroecological research can complement the small-scale experiments. It can place them in a larger perspective that explicitly considers spatial and temporal variation, the openness of ecological systems to exchanges across their boundaries, and the influence of history.

In theory, the effects of spatial and temporal variation can be assessed by performing replicated experiments at many different sites and by continuing them for long periods. In practice, this is very seldom done, because it is difficult to obtain resources and to maintain the commitment to such a massive research program. Compilation and analysis of nonexperimental data offers a practical alternative. Questions such as, how typical a year was 1988? or, how representative is my research site? can often be addressed using data on abiotic environmental conditions and species composition that are already available.

For example, when we obtained different results from our rodent and ant competition experiments at Sonoran and Chihuahuan Desert sites, I began to worry about the generality of the results from our experiments on competition among rodents. These experiments have been performed at only one Chihuahuan Desert site. One obvious question is how representative is the assemblage of rodent species that occurs at this site. Margaret Kurzius and I compiled data on the species composition (presence/absence) of granivorous rodents at 202 sites of relatively uniform desert habitat throughout the arid region of the southwestern United States (Brown and Kurzius 1987, 1989). The results were very informative. First, our Chihuahuan Desert experimental site had a combination of species not duplicated at any other site. Obviously the species composition at this site is not typical of any extensive region. Second, the 202 sites were inhabited by 136 different combinations of 29 species (see fig. 3.2). No single combination (of two species) occurred at more than 6 sites. Therefore, there is really no such thing as a typical desert rodent assemblage from the standpoint of species composition. It is certainly not clear to what extent our experimental results, or the results of any other experimental study of competition among desert rodent species performed at only one site, represents a general outcome. The most that we can hope for is that some aspects of the interactions among rodent species are much more general than the identity of the species themselves. It remains to be seen whether this is true.

Many experimental studies in ecology treat the systems, at least implicitly, as if they were closed. Sometimes this is done by choosing naturally isolated experimental units, such as ponds, where inputs of materials and organisms are minimal. Sometimes it is done by using artificial barriers or periodic removal to maintain specified conditions. Most often it is done by limiting the duration of the experiment so that the influence of extrinsic factors is minimized. There are good reasons for using relatively closed systems. Movements of materials and organisms across the boundaries may be difficult to monitor, and may have effects that are difficult to interpret.

Most ecological systems are not closed, however, and the use of closed systems for experiments may bias these studies against detecting important, larger-scale phenomena. In our desert experiments, we have deliberately tried to maintain "openness"; although we use fences to maintain the exclusion of selected species, we have designed them so as not to interfere with the movements of the remaining species. This design, coupled with long-term monitoring, has enabled us to appreciate just how open the system is. Thus, during the 15 years of the study we have docu-mented multiple local colonization and extinction events in several rodent species (Brown and Heske 1990b). We also found that several rodent spe-cies, characteristic of grassland habitats and never captured on the site during the first 10 years, subsequently colonized those plots where re-moval of kangaroo rats had caused dramatic increases in grass cover (Brown and Heske 1990a; Heske, Brown, and Mistry 1994). These results show how the response of complex ecological systems to environmental change depends upon their openness, and on the roles of colonization and extinction in mediating the interactions between local communities and the regional assemblages in which they are embedded (for other ex-amples, see Ricklefs and Schluter 1993).

Yet another example is provided by the work of Roughgarden and col-leagues on intertidal communities. There is a long history of experimental studies on rocky shores. The vast majority have been conducted for only a few weeks or months, on one stretch of shore, on experimental plots less than one square meter in area. These experiments have repeatedly demonstrated the effects of local abiotic conditions, especially exposure between tides, and biotic interactions, especially predation and competi-tion, on the abundance and distribution of species (e.g., Connell 1961a,b, 1975; Paine 1966; Lubchenco and Menge 1978). These kinds of experi-ments were successful at accounting for the outcome of ecological inter-actions within a season once colonization had established the initial spe-cies composition of the community.

Roughgarden noticed, however, that there was great year-to-year varia-tion in the initial species composition of the communities on the shore of Monterey Bay. These initial differences could not be explained in terms of the interactions within the local intertidal habitat, but they did affect the outcomes. Roughgarden and colleagues focused their research on barnacles, which varied greatly from year to year, but when abundant strongly affected community composition by dominating competition for space (Roughgarden, Gaines, and Pacala 1987; Roughgarden, Gaines, and Possingham 1988). They showed that the initial densities of barnacles were determined by the rate at which their planktonic larvae settled on the rocks. Larval settlement rate depended on interannual shifts in the ocean currents that dispersed the relatively short-lived larvae. From satel-lite images showing the configuration of the currents, the investigators were able to account for the initial densities of barnacles and thus the final composition of the intertidal communities.

These results are illuminating in two ways. First, they show the in-creased understanding that can come from a macroscopic approach that tries to measure and account for, rather than eliminate or ignore, the exchanges across the boundaries of local ecological systems. A second and more general point is that differences in initial conditions can dramatically alter the outcome of ecological interactions. This sensitivity to differences in initial conditions is a general characteristic of complex nonlinear systems (see chap. 2). It means that historical events play a major role in ecology. Seemingly insignificant changes in abiotic conditions or species composition that happened long ago or far away can have large, irreversible effects on the structure and dynamics of local ecological systems.

ENERGETICS AS A UNIFYING THEME

The Two Currencies: Energy and Offspring

One problem with any attempt to unify the various subdisciplines of ecology is that they use different currencies. On the one hand, most ecologists working at the level of individual organisms use rates of energy exchange, dE/dt, as the basic currency in their theoretical and empirical studies of responses to physical stress, strategies of foraging in animals and resource acquisition in plants, and tactics of space use and territorial defense. Even though the proximal currency may sometimes be nitrogen, protein, or even predation risk, rather than energy per se, thermodynamic considerations are paramount. Individual organisms are viewed as maximizing their fitness by acquiring scarce resources from the environment, using them to maintain the homeostasis of the individual, and allocating them to offspring. Also, at the level of ecosystems, many ecologists are concerned with the distribution and exchanges of either energy itself or rare materials, such as nitrogen or phosphorus, that are required and concentrated by organisms.

In between the levels of individual organisms and entire ecosystems are populations and communities. The evolutionary ecologists who study these systems virtually all use some version of changes in the numbers of individuals, dN/dt, as their fundamental currency. This can be attributed to the evolutionary background of most population and community ecologists. Population geneticists and microevolutionists measure evolution in terms of the rate of change in the frequency of an allele or the distribution of a quantitative trait in a population. This is formally equivalent to measuring the differential survival and reproduction of individuals with alternative genotypes or phenotypes.

It will be difficult to unify the various subdisciplines of ecology and to begin the long-awaited synthesis between ecology and microevolution so long as the disciplines use different currencies. What are the prospects? I believe that the current use of different currencies and the resistance to change are largely reflections of different intellectual histories. Elsewhere I (Brown 1981; see also Ulanowicz 1989) have attributed the focus on energy (dE/dt) to the influence of Raymond Lindemann (1942), Eugene Odum (1959, 1969), and H. T. Odum (1983), and the use of a biological fitness currency (dN/dt) primarily to Robert MacArthur (1960a, 1961, 1962, 1965). The perpetuation of this historical dichotomy is reflected in the training of contemporary ecologists. Most ecologists who work at the individual level have backgrounds in physiology, other areas of organismal biology, and the physical sciences. Many ecosystem ecologists have strong training in the physical sciences. In contrast, population and community ecologists tend to have backgrounds in evolutionary biology and population genetics.

The Maximum Power Principle

My own view is that dE/dt and dN/dt are not fundamentally different (Brown 1981, in press-b; see also DeAngelis, in press; Wiegert, in press). Both survival and reproduction are thermodynamic phenomena. In order for an individual organism to maintain homeostasis and produce offspring, it must acquire rare materials from the environment, it must maintain them against an organism-environment concentration gradient and against other organisms seeking to exploit them, and it must package them into offspring that have the capacity to grow and develop into new individuals. To accomplish these tasks, the organism must take in and transform energy to perform the work of contracting muscles, conducting nerve impulses, and moving materials across membranes against concentration gradients—and ultimately such complex activities as interacting with other species. In a physical sense, the individual organism is a complex, open system that maintains itself far from thermodynamic equilibrium and reproduces itself by the continual throughput of energy that is degraded to perform work (Lotka 1922, 1925; Schrodinger 1946; Odum 1983; Schneider and Kay, in press; Brown 1994, in press).

The conversion of energy into biological work is the essence of fitness. Thus, it is theoretically possible to express dN/dt in terms of dE/dt or some similar thermodynamic currency. It should also be possible to do this in practice. Physiological and behavioral ecologists do so when they use energy acquisition and allocation in their theoretical and empirical studies of foraging and reproduction. It should be possible to develop similar conceptual frameworks for investigating life history, adaptive evolutionary change, and coevolution of interacting species. A redefinition of fitness in physical terms would avoid the troubling logical dilemmas of potential circularity and tautology that have bedeviled evolutionary biologists ever since Darwin (e.g., Peters 1976, 1991).

This is not so revolutionary as it may seem. Several physicists have been making essentially this same argument at least since the turn of the century. Thus, Boltzmann (1905) said that "[the] struggle for existence is a struggle for free energy available for work." Similarly, Schrodinger (1946) pointed out that "organization [is] maintained by extracting 'order' from the environment." A few ecologists and evolutionary biologists, most notably Lotka (1922, 1925), H. T. Odum (1983; Odum and Pinkerton 1955), and Van Valen (1973a, 1976) have tried to make more explicit the relationship between energetics and fitness.

One such effort is the model of Brown, Marquet, and Taper (1993) presented in chapter 5. This is a restatement, in more explicit biological terms, of the "maximum power principle" of Lotka (1922, 1925) and Odum (1983; Odum and Pinkerton 1955). According to this principle, it is power, the rate of transformation of energy into work—and not efficiency, the ratio of transformed energy output to gross energy input—that determines success, fitness, and dominance in most competitive, energy-limited systems, including physical machines, organisms, and human economies. Many ecologists and evolutionary biologists have incorrectly assumed that natural selection tends to increase efficiency. If this were true, to give just one example, endothermy could never have evolved. Endothermic birds and mammals are extremely inefficient compared with both their ectothermic reptilian ancestors and contemporary reptiles and amphibians. Endotherms expend energy at high rates in order to maintain a high, constant body temperature and the associated capacity for high levels of activity independent of environmental temperature. This enables birds and mammals to occur in many environments where reptiles and amphibians cannot, and to dominate resource use in virtually all environments where they coexist with ectothermic vertebrates (Turner 1970).

By defining fitness as reproductive power, dW/dt, the rate at which energy can be transformed into work to produce offspring, our model makes a major conceptual leap (see also Van Valen 1976; Reiss 1989). Admittedly, the model is in a very early stage of development. It needs to be refined, applied to specific problems of physiological ecology, life history evolution, species packing, and macroevolution, and tested with appropriate data. Even if the model is ultimately changed or totally discarded, however, it focuses attention on the intimate relationship between two fundamental aspects of living systems, energetics and allometry.

Energetics and Allometry

Perhaps one of the reasons that population and community ecologists have generally not used energetics as a currency is that it has been difficult to measure energy intake and expenditure by different kinds of organisms for different kinds of activities. These problems have now largely been surmounted by comparative physiologists and physiological ecologists, who are using isotopes (such as doubly labeled water), telemetry, and other techniques to measure the costs of reproduction, locomotion, and other activities in free-living individuals in the field (e.g., Nagy 1987). Data for a large number of organisms are increasingly available (e.g., summarized in Peters 1983; Calder 1984; Schmidt-Nielsen 1984; Bonner 1988; Reiss 1989).

More importantly, analyses of these data are revealing robust relationships between body size and virtually all characteristics of the organisms. Many morphological and physiological traits (e.g., brain size, gut capacity, metabolic rate, heart rate) appear to scale monotonically with size and to be well characterized by allometric power functions. These allometric equations often explain a large proportion of the variation, so they can be used to predict values for species that have not been measured. It is possible to take advantage of two additional features of these allometric relationships to increase the accuracy of such predictions. First, closely related organisms characteristically exhibit less variation about fitted allometric regressions than more distantly related ones. Second, body size and its correlated traits often vary by several orders of magnitude, while the deviations around the regression lines are relatively small. So considerable accuracy is possible if extrapolations are restricted to fairly closely related species that differ substantially in size.

Variation in other traits, especially behavioral and ecological ones (e.g., running or flying speed, territory or home range size, population density and growth rate), is apparently not so tightly constrained by size. There can be great variation among different species, and sometimes different populations of the same species, of similar body size. Furthermore, the relationships may be hump-shaped (or U-shaped), exhibiting maximum (or minimum) values of the parameter at some intermediate size. As I have suggested in chapters 5 and 7, variation in many of these traits would be better described by constraint envelopes than by allometric equations. In these cases, it is hazardous to predict values for populations or species that have not been studied. There is still much to be learned, however, from the shapes of the constraint envelopes.

A body of theory to explain the empirical patterns of size-dependent variation is beginning to be developed. There is still a long way to go,[1] but

---

1. Perhaps the most vexing problem of all is why metabolic rate should scale as approximately $M^{0.75}$. There are several hypotheses (e.g., see McMahon 1973; Calder 1984; Schmidt-Nielsen 1984), but I am not convinced that any of them are sufficiently general to account

testable hypotheses are being proposed, and appropriate data to evaluate them are being collected and analyzed. What I find most exciting about these hypotheses is their potential to synthesize information about two of the most fundamental characteristics of living things: (1) their more than twenty orders of magnitude variation in body size and (2) their diverse but constrained mechanisms for acquiring energy and essential materials from their environments and allocating them to growth, survival, and reproduction. Such a synthesis would contribute to our understanding of how the morphology, physiology, and behavior of individual organisms influence the structure and dynamics of populations, communities, and large-scale biotas, and vice versa. It would contribute to our understanding of the energetic and thermodynamic basis of fitness.

In pursuing the ideas of Gause, Hutchinson, and MacArthur, many subsequent evolutionary ecologists focused on the question of how similar species could be and still coexist. They performed many comparative studies of species that were closely related and very similar in size and other characteristics.[2] While this may have been understandable, it ignored one of the most obvious and important features of the natural world: most of the species that occur together on any spatial and temporal scale are different in size and not very closely related. Think of the five to seven orders of magnitude variation in the body sizes of fishes, reptiles, birds, and mammals, not to mention the much greater differences between these vertebrates and the arthropods and other invertebrates. I suggest that there is much to be learned from asking how the current coexistence and the historical evolutionary diversification of these organisms is influenced by the differences among them in size and other characteristics. I suggest that one way to approach this question is a macroecological approach that focuses on energetics and allometry.

CONCLUDING REMARKS

One contribution of the macroecological research program is its ability to present the big picture. It can place the results of the predominantly small-scale, short-term ecological experiments in a larger perspective. It can shed light upon the question of which experimental results are specific to a particular time, place, organism, or habitat, and which are more general. It can emphasize the openness of small-scale systems to ex-

---

for a relationship that holds for organisms ranging from microbes to mammals, and from aquatic to terrestrial forms.

2. Aside from the desire to pursue then-topical ideas about competitive exclusion and limiting similarity, one motivation for such studies was the "experimental design" concept that tries to hold most variables constant or nearly so.

changes of organisms and materials across their often arbitrary boundaries.

By focusing on the major features of diversity, macroecological studies have already identified interesting patterns of relationships among variables. It remains, however, to evaluate the generality of the patterns across different kinds of organisms and environments, and to develop and test mechanistic hypotheses to account for the relationships. While a great deal of research remains to be done, the signs are encouraging. Many of the patterns appear to be very general.

More importantly, however, a mechanistic perspective based on energetics, allometry, and their interrelationships has the potential to explain many of the patterns. This is fortunate, because one of the most severe limitations of macroecology is the few kinds of data that are available in sufficient quantity for analysis of statistical patterns. Because energetics is so fundamental, and because the influence of body size on aspects of organismal structure and function, including energetics, is so universal, it may be possible to explain a great deal of ecology and evolution with some general principles and a modest amount of information. Particularly exciting is the prospect that fitness might be defined in energetic terms, providing a common currency to unify many areas of ecology and evolutionary biology.

11 Synthesis

Biogeographic and Macroevolutionary Implications

BIOGEOGRAPHY

The Ecological Limits on the Geographic Range

The boundaries of the geographic range of a species are determined by ecological limits: by an interaction between the niche requirements of the organisms and the abiotic and biotic characteristics of the environment. The only exceptions are the few species, now primarily recently introduced exotics, that are temporarily not limited and are expanding their ranges exponentially (e.g., Brown and Gibson 1983; van den Bosch, Hengeveld, and Metz 1992).

Earlier (chaps. 3 and 6) I pointed out that the boundary of the range tends to occur where environmental conditions cause the sum of death plus emigration rates to exceed the sum of birth plus immigration rates. It is important to take emigration and immigration rates into account because of metapopulation dynamics. At least in mobile organisms, some peripheral populations may be sinks, sustained by immigration even though death rates exceed birth rates (e.g., Grinnell 1922; Hanski 1982c; Holt 1983; Pulliam 1988). Also, even in the absence of immigration, some peripheral populations can persist for long periods with little or no recruitment if the survival rate of established individuals is high (e.g., Stevens and Fox 1991).

Three other points about the ecology of range boundaries are relevant. First, they are dynamic (e.g., Hengeveld 1990). If the limiting conditions vary, then the range boundaries will fluctuate in space and time. The rate and magnitude of the shifts will depend on both the pattern of environmental variation and the population dynamics of the organisms.

The second point is that the environmental conditions that set the boundaries of the range also define what biogeographers call barriers: the inhospitable areas that cannot sustain a population, so that dispersal across them must usually occur rapidly and during some resistant stage of the life cycle. The recognition of barriers implies that they separate areas that are at least potentially habitable. The success of exotics in becoming established in new regions after human assistance in crossing barriers shows that indeed many species do not presently occur in all regions where they could sustain viable populations. Sometimes widely separated areas are actually inhabited, however, and we infer from the disjunct populations and environmental history either (1) that even though the barrier was present, at least one propagule successfully crossed it in the past (what biogeographers would call the dispersalist hypothesis), or (2) that the barrier was absent in the past because environmental conditions were different and the organisms could live in the intervening areas (the vicariance hypothesis).

The third point is that all range boundaries are ultimately set by abiotic conditions—by the variation in physical conditions across the earth's surface. Proximate limits may indeed be set by some biotic interaction, such as the presence of a competitor, predator, or pathogen, or by the absence of an essential prey, host, or mutualist. But the reason why the other species has its limiting effect in one particular place and not another must ultimately be due to some characteristic of the present or past abiotic environment.

The Meaning of History

Unfortunately, biogeographers use the term "history" to refer to two different phenomena, which I will call the history of place and the history of lineage. The history of place is the sequence of changes in past environments. It is historical geology, geography, and ecology. Events at all scales, from the tectonic movement of continental plates to the appearance of new volcanic islands to small changes in climate and human land use, have left their legacy on the present distribution and diversity of organisms. The magnitude and effects of these changes can be difficult to grasp. Just 20,000 years ago a vast ice sheet tens of meters thick covered most of Canada and the northernmost United States, and enormous lakes filled what are now desert valleys in the southwestern United States. When the first European settlers arrived on the eastern shore of North America only 400 years ago, they found extensive forests containing chestnut trees, passenger pigeons, ivory-billed woodpeckers, elk, wolves, and mountain lions.

By the history of lineage I mean the sequence of speciation and extinction events that have occurred since descent from a common ancestor. These are the events that are documented in phylogenetic (cladistic) reconstructions of evolutionary relationships. Historical biogeographers have obtained valuable insights by mapping such phylogenetic reconstructions or cladograms onto geographic space and comparing the resulting area cladograms for different lineages (e.g., Humphries and Parenti 1986; Myers and Giller 1989; Brooks and McLennan 1991).

The principle of such cladistic biogeographic reconstructions is straightforward. There was only one history of the earth, and the sequence of events in a certain place should have affected the distributions of all organisms that were living there at the time. The same kinds of barriers that limited distributions and led to speciation in one lineage should have had similar effects on other lineages. Consequently, area cladograms are predicted to be congruent among different lineages (clades), and the history of lineage should reflect the history of place (Croizat 1958; Nelson and Platnick 1981; Humphries and Parenti 1986; Brooks and McLennan 1991). Perhaps the most convincing examples are the apparently congruent geographic distributions and phylogenetic relationships of certain parasites and their hosts (e.g., Brooks 1985; Brooks and McLennan 1991 and included references). These are taken as evidence of cospeciation, simultaneous speciation in two or more lineages as a result of the same historical geologic or ecological event.

While I do not doubt that there are some bona fide cases of congruent lineages that reflect a common biogeographic history of speciation and dispersal events, I remain skeptical about the extent to which the reconstructed history of lineage can be used to reconstruct the history of place. I have two main concerns.

First, phylogenetic reconstructions are almost always incomplete. They include only the branches represented by contemporary species. They rarely incorporate evidence from fossils, and even if they did, they would still miss many of the extinction, dispersal, and speciation events that actually occurred. This missing information is not so critical in phylogenetic reconstruction, because the relationships among contemporary forms can be inferred without a complete knowledge of all past speciation and extinction events. It is much more critical in biogeographic reconstruction, however, because the algorithms of cladistic biogeography assume a faithful mapping of the complete histories of lineage onto the history of place.

My second concern has to do with the way that organisms have responded to the changes in geology, climate, and other environmental conditions. As indicated above, some of these changes have been rapid and geographically extensive. As indicated in chapter 8, the response of contemporary organisms, even closely related and ecologically similar species, to environmental change in space and time is often individualistic and difficult to predict. Finally, we know that plant and animal species have exhibited enormous dispersal and shifts in their geographic ranges in just the last 20,000 years (e.g., Bernabo and Webb 1977; Davis 1986; Graham 1986; Betancourt, Van Devender, and Martin 1990). Taken together, these facts cast doubt on the assumption that the history of lineage and the history of place are congruent. The geographic ranges of the species in many groups have probably experienced so many large shifts since the speciation events that it would be difficult or impossible to reconstruct the spatial pattern of the history of lineage.

Having voiced my skepticism, let me add two comments. First, I would be delighted to be proved wrong. In fact, I am optimistic that the approaches of phylogenetic reconstruction and cladistic biogeography can be combined with those of paleontology and paleoecology to reveal a great deal more about the history of lineage, the history of place, and their relationships to each other. I cannot stress enough, however, the importance of including the direct insights into the nature of past environments and past distributions of organisms that can only come from the fossil record.

Second, I think the distinction that I make between the history of place and the history of lineage is a useful one that can ultimately contribute to a synthesis of historical and ecological biogeography. The history of place is an environmental history. We can infer that previous environmental conditions limited the distributions of organisms in the past by processes similar to those operating today to limit the distributions of their descendants. To a large extent the histories of places can be viewed as independent variables and the historical distributions of organisms can be viewed as dependent or response variables. As pointed out in chapters 2 and 6, the physical template of the earth, and in particular its geology, climate, and oceanography, ultimately determine the distributions of living things. In important ways, however, the activities of plants, animals, and microbes feed back to alter the environment and affect the distributions of other organisms. To give one example, the distributions of many parasites may be closely tied to the distributions of their hosts, which provide many of the essential environmental conditions necessary for their survival and reproduction. Even more obvious is the enormous effect of the many activities of our own species on the distributions of many other organisms. I will consider the consequences of some of these human influences in the next chapter.

MACROEVOLUTION

The Macroevolutionary Research Program

In the 1970s some paleontologists coined the term "macroevolution" for the large-scale, long-term patterns and processes in the evolution of lineages revealed in the fossil record (e.g., Eldredge and Gould 1972; Gould and Eldredge 1977; Eldredge 1979, 1985, 1989; Stanley 1979, 1982; Gould 1980). Especially in its infancy, the macroevolutionary research program was centered around two concepts: (1) "punctuated equilibrium," a term used to characterize long periods of virtual stasis interspersed with short periods of rapid change often apparently associated with speciation events; and (2) "species selection," a term used to refer to differential rates of speciation or extinction that resulted in increases or decreases in the species diversity of different lineages over time.

I will make only two points about punctuated equilibrium. First, the relative stasis that is observed in the characteristics of species over long time periods in the fossil record seems likely to be easily explained by stabilizing selection. Not only may the environmental factors that constitute the niche dimensions of species remain relatively constant over substantial periods, but even when the environment changes, the species may respond by shifting their geographic ranges to remain in a similar environmental setting rather than remaining in situ, responding to directional selection, and changing substantially. Clearly this is what happened in North America during the Pleistocene: during the glacial periods, species of trees, mammals, and other organisms virtually identical to contemporary forms occurred hundreds of kilometers outside their current range boundaries (e.g., Davis 1986; Graham 1986; Betancourt, Van Devender, and Martin 1990).

Second, both the apparent stasis and the rapid changes associated with speciation events might be explained in large part by the interspecific relationship between abundance and distribution discussed in chapter 4 and associated biases in the fossil record (see also Lazell 1966; Brown 1984). Thus, the widespread and abundant species should predominate in fossil samples just as they do in samples of contemporary communities. These species should be especially subject to stabilizing selection; there is little scope for them to improve their ecological performance, because they already are tolerant of wide ranges of abiotic conditions and able to coexist with many other species and to use a wide variety of resources. The rare and restricted relatives of these species, in contrast, should be poorly represented by the sampling of the fossil record. They also should have considerable scope to improve their ecological performance; they should increase rapidly in both geographic range and local abundance if either (1) the environment changes to favor a species with their niche requirements, or (2) they come up with an evolutionary innovation that enables them to circumvent some previous environmental limitation. In either case, their "sudden" appearance in the fossil record as an abundant, widespread species will tend to appear to be the result of a speciation event.

The concept of species selection is intimately related to what I call species dynamics (chap. 9). A process analogous to natural selection operating at the level of individuals almost certainly operates at the level of species. Darwin (1859) first pointed out that four conditions are both necessary and sufficient for evolutionary change: (1) units reproduce themselves, potentially producing more descendants than can survive; (2) there is variation among units; (3) at least some portion of this variation is heritable; and (4) there is differential survival and/or reproduction of the units possessing different heritable characteristics. If these conditions are met, then descent with modification is inevitable. Further, if the variants survive and proliferate differentially based on their success in coping with some aspect of their environment, then there must be descent with adaptive modification—and we have a process analogous to natural selection.

It seems increasingly clear that species meet Darwin's four criteria for a process analogous to natural selection (e.g., Stanley 1979; Sober 1984; Eldredge 1985, 1989; Wilson and Sober 1989; Williams 1992; T. Nusbaum, J. H. Brown, and M. L. Taper, unpub.). Species reproduce themselves by speciation, potentially producing more descendant species than can survive (criterion 1). There is variation among species in many traits (criterion 2). And there is abundant evidence of differential speciation and/or extinction of species or lineages of multiple species with different characteristics (criterion 4: e.g., Doyle and Donoghue 1986; Jablonski 1986a,b; Dial and Marzluff 1989; Donoghue 1989; Brooks and McLennan 1991; Taylor and Gotelli, in press).

A critical question, then, is whether a significant portion of the variation among species is heritable in a sense analogous to genetic heritability (criterion 3). Jablonski (1987) used the equivalent of parent-offspring regression analysis in quantitative genetics to analyze the heritability of area of geographic range in fossil marine mollusks. T. Nusbaum, M. L. Taper, and I (unpub.) used the equivalent of sib-sib analysis to assess heritability at the species level of area of geographic range and individual body size in contemporary North American birds and mammals. Both traits are highly heritable, and Jablonski's and our analyses give surprisingly similar values for heritability of area of geographic range (table 11.1).

At the level of individuals within populations there are three major

TABLE 11.1. Species-level heritability of two traits, average adult body mass and area of geographic range, of North American terrestrial mammals and land birds.

Class	Heritability of	
	Mean body mass	Area of geographic range
Mammals		
Genus	0.96	0.30
Subgenus	0.97	0.32
Birds		
Genus	0.96	0.59

Source: T. Nusbaum, J. Brown, and M. Taper, unpub.
Note: Heritability was estimated using one-way ANOVA to calculate the proportion of variation (r^2) in each class of vertebrate that can be attributed to the distribution of species among genera or subgenera. All values are significant at $P < .005$.

evolutionary forces: drift, migration, and selection. All have their ana-logues at the level of species within biotas. Since species satisfy Darwin's four criteria, random speciation, extinction, and geographic range shift events will result in processes at the species level analogous to drift and migration. But such random processes are, by definition, not selective. That is why I prefer to use the neutral terms species dynamics and macro-evolution to refer to changes in diversity when the causal mechanisms have not been identified. As I have shown in chapter 9, however, in many cases speciation, extinction, and geographic range shifts are not random, but are influenced by the relationships between organisms and the envi-ronment. Thus, species selection is a major force in the evolution of bi-otic diversity.

"Emergent" Traits and Species Selection

Recently, some macroevolutionists (e.g., Vrba 1983; Vrba and Eldredge 1984; Eldredge 1985) have suggested that a trait must be "emergent," or expressed only at the species level, in order to be subject to species selec-tion. Thus, while they might grant that area of geographic range (or an insular as opposed to a continental distribution) might be such an emer-gent trait, body size or habitat affinity could not be, because these latter characteristics are expressed and are subject to selection at the lower level of the individual. Differential speciation or extinction based on traits ex-pressed at the individual level has been referred to as "species sorting" rather than species selection.

I think that use of the term "emergent" in this sense is unnecessarily restrictive and obfuscates rather than clarifies the process of species selec-tion. My position is based on two considerations. First, I prefer a more operational use of the term "emergent" (see chaps. 2 and 3) to describe any property of a higher level of organization that is reflected in the statistical distribution of some measured attribute for a large number of equivalent, lower-level units. Thus, for example, a species has an emergent body size that can be characterized in terms of the distribution of body sizes among its component individuals, and can be measured in terms of the quantitative characteristics of that distribution (e.g., mean, variance, skewness, kurtosis, etc.). This operational way of defining "emergent" avoids the logical difficulties of trying to decide when a trait is simply a collective property of an assemblage of lower-level units and when it is some special property of the higher level alone. Indeed, the theories of hierarchies and complex systems would question whether there is any higher-level trait that is not in some sense a collective expression of the attributes of the lower-level units. Consider a spectrum of characteristics measured at the species level: mean body mass, sex ratio, age structure, variance in male reproductive success, average population density, area of geographic range, time since divergence from most closely related species. Which ones are emergent? By my operational definition, all of them are.[1]

Defining an emergent trait in this operational way helps to make my second point, which is that whether selection acts on a trait at any level depends not on whether the trait is expressed only at that level, but on whether the trait causes differential survival and proliferation of the units at that level (see also Sober 1984; Williams 1992). Thus, for example, selection can act on cancers both at the cellular level and at the individual level; in this case the two levels of selection will tend to act in opposition to each other, because the unregulated proliferation of certain cell lineages ultimately reduces the survival and reproduction of the individuals with the cancerous cells. Similarly, species selection can act on traits that cause differential survival and/or reproduction of individuals so long as these traits also affect speciation and extinction.

Consider the case of body size. Imagine two cases in which large size suddenly becomes highly disadvantageous, perhaps because a meteorite impact causes global cooling and a reduced food supply. In the first case, 90% mortality selectively eliminates the largest individuals within a species of relatively large size. The result is that the species persists, but if size is a heritable trait, the distribution of body sizes for some succeeding generations is shifted downward. This is individual selection, because it

---

1. If others are unwilling to use this broad, operational definition of "emergent," then I suggest that we need a new term for the statistical properties of assemblages of many equivalent units. I do not recommend this, however, because I do not believe in the proliferation of jargon, and I think that the existing term "emergent" is difficult to define rigorously in any other way.

occurs through the differential survival and/or reproduction of individuals within populations. In the second case, large size is so disadvantageous that all individuals in the species are killed. This is species selection, because the entire species lineage is the unit that differentially survives and/or reproduces. It is not simply individual selection writ large. In fact, it is immaterial whether the larger individuals within the species are somewhat more susceptible and die first. What is important is that the emergent property, the distribution of sizes among the individuals of the species, causes the entire species to go extinct. Once a species lineage has been terminated by extinction or produced by speciation, a qualitatively different level of biological organization has been subjected to and affected by the evolutionary process.

Stebbins and Ayala (1981; see also Charlesworth, Lande, and Slatkin 1982) criticized the concept of species selection by suggesting that, even if it sometimes occurs, it is a weak force because it acts much less frequently than individual selection. It is true that individual selection can potentially operate every generation, whereas the operation of species selection will often be apparent only on the scale of the millennia required for changes in species composition to be observed in the fossil record.[2] But this does not mean that species selection is a weak force. By causing the termination or origination of entire phylogenetic lineages, species selection, when it occurs, has a much larger effect on organic diversity than the death or reproduction of a single individual.[3] Certainly, the selective extinction of the giant reptiles at the end of the Cretaceous was every bit as important an evolutionary event as the evolution of industrial melanism in the peppered moth!

MACROECOLOGY AND THE ENVIRONMENTAL FILTER

Evidence for Species Selection in the Fossil Record

It is important to make the distinction between all evolution at the species level and the subset that occurs as a result of species-level selection. In

2. This is not always true, however. The meteorite impact that eliminated the dinosaurs and several other major lineages at the end of the Cretaceous was a sudden event that probably accomplished most of its species-level selection in less than one generation. Another example, developed in the next chapter, is the wave of extinctions during the last several thousand years that is directly attributable to humans, which has been highly selective with respect to traits such as body size, and which has occurred so rapidly that there is no evidence of simultaneous individual-level selection for smaller size within these species.

3. George Williams (1992) reaches a similar conclusion. For reasons given in chapter 3, however, I disagree with Williams's argument that species selection is just a special case of what he calls clade selection.

much of the early macroevolutionary literature, the differential proliferation of some lineages and the dwindling of others that were observed in
the fossil record were attributed to stochastic sampling processes analogous to drift rather than to selection. For example, Raup et al. (1983;
Raup and Gould 1984; Gould et al. 1977) showed that computer simulations of the evolution of lineages that incorporated random probabilities
of speciation and extinction could produce distributions of traits among
species over time that were similar to observed distributions. Similarly,
Gould (1988; but see Stanley 1973; Dial and Marzluff 1989; Maurer,
Brown, and Rusler 1992) suggested that the diversification of traits such
as body size among species over time could be explained as a result of the
random increase in variance that would inevitably occur when multiple
speciation events gave rise to descendant species that differed somewhat
from their ancestors. There was a similar tendency to view the extinctions
detectable in the fossil record as the consequence of essentially random
processes.

As paleobiologists have compiled larger and more accurate data sets
and subjected them to more rigorous analyses, however, they have found
that many of the speciation, extinction, and colonization events recorded
in the fossil record were nonrandom. Thus, there is increasing evidence
for species selection: a process analogous to natural selection but operating at the level of species within biotas. Furthermore, both paleobiologists and neontologists have identified traits that are associated with the
differential success or failure of lineages. For example, Boucot (1976; see
also Dial and Marzluff 1989) suggests that a recurring pattern in the fossil
record is that lineages of small-sized marine invertebrates exhibit higher
rates of speciation than clades of large-bodied forms. The explosive speciation of cichlid fishes, especially in the Great Lakes of central Africa, has
been attributed to innovations in the structure and biomechanics of the
pharyngeal jaws that have enabled the species to use an amazing variety
of food types and feeding modes (Fryer and Iles 1972; Liem 1973). Graham and Lundelius (1984) and Owen-Smith (1988) demonstrate the differential extinction of large mammals in North America, presumably
caused at least in part by human colonization and hunting, during the last
20,000 years (fig. 11.1). Jablonski (1986a,b; see also Hansen 1980; Jablonski and Lutz 1983) shows that fossil mollusk taxa with certain larval characteristics speciated at different rates and differentially survived or disappeared during periods of both mass extinction and background extinction.
Simpson (1980), Marshall et al. (1982), Reig (1989), and others describe
the greater geographic range expansion and speciation of North American
mammal lineages in South America than the converse after the connec-

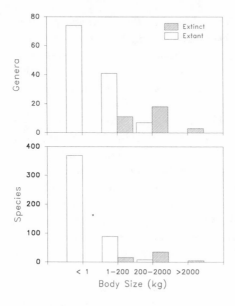

FIGURE 11.1. Frequency distributions of late Pleistocene North American land mammal genera and species as a function of four approximately logarithmic classes of body mass, comparing the body sizes of those taxa that went extinct with those that survived to the present. The differential extinction of large forms is clear. (Compiled from various sources, but data on the sizes of extinct mammals come largely from Martin and Klein 1984.)

tion of the two continents by the formation of the Isthmus of Panama about 3 million years ago.

Whenever species selection in the narrow sense occurs, the environment is the filter that determines the probabilities of speciation, geographic range shift, and extinction. A major synthesis between macroecology and macroevolution, therefore, will come from elucidating the interactions between species and environmental conditions that are responsible for the differential success of lineages.

Mechanisms

It is one thing to document the differential survival and/or proliferation of species with particular characteristics; it is another matter to determine the mechanisms of such species selection. This problem is not unique to selection at the species level. At the level of individuals within natural populations it is also much easier to document a change in gene frequency or a quantitative character than it is to understand how the process of selection has affected survival, reproduction, and migration to cause the change. At both the species and individual levels, much of the difficulty can be attributed to the fact that the units of selection comprise many traits, which exhibit complex patterns of correlated variation and have multiple effects on interactions with the environment. Body size is a prime example. For reasons mentioned in chapters 5, 6, 7, and 9, body size influences many attributes of both individuals and species that may in turn affect probabilities both of survival and reproduction and of speciation and extinction.

I think it is possible, however, to begin to understand some of the important mechanisms of species selection. Paleobiologists (e.g., Sepkoski 1982; Jablonski 1985, 1986a,b, 1987; Raup 1986; Raup and Jablonski 1986) have shed a great deal of light on the process and consequences of the extinctions recorded in fossils. As mentioned in chapter 9, these extinctions have often been divided into two categories: background and mass extinctions. Jablonski's studies of fossil mollusks suggest that different kinds of species tend to have survived during the two kinds of extinctions: having a large geographic range is associated with higher probability of survival during times of background extinctions, while belonging to a geographically widespread clade (but not necessarily being a species with a large geographic range) tends to characterize the survivors of mass extinction events. This pattern suggests that small geographic ranges and chronically low total population sizes (see chap. 6) do indeed increase the probability of extinction during times of normally fluctuating environments. But large ranges and populations offer little protection from the extreme environments that cause mass extinctions. Then some members of the most speciose lineages tend to survive, perhaps just by chance, because those clades have more species to begin with, or perhaps because those clades have a greater probability of containing one or more species that have the special attributes required to tolerate the unprecedented environmental change.

So far, most of my discussion of species selection has focused on extinction. It is not difficult to document the selectivity of extinction in many lineages, and, as indicated above, we are beginning to obtain the information needed to understand the underlying mechanisms. Much less has been said, both above and in the literature, about the selectivity of speciation.

Differential speciation is potentially also a powerful component of species selection. There are many examples in the fossil record and in the diversity of contemporary clades of the spectacular diversification of certain lineages. These include the domination of insular faunas owing to radiations of finches in the Galápagos, honeycreepers in the Hawaiian Islands, and cichlid fishes in the Great Lakes of central Africa. Two of the most spectacular radiations recorded in the fossil record are the rapid diversifications of angiosperm plants and teleost fishes. It is probably not owing simply to chance that certain lineages have, for a substantial period of time, produced more species than they have lost through extinction. It is difficult, however, to identify the causes for their success. One pattern that seems clear, however, is that clades of smaller organisms have tended to speciate and to leave more surviving descendants than their larger relatives (e.g., Boucot 1976; Dial and Marzluff 1988, 1989; Martin 1992).

One example of such explosive speciation is the radiation of the mouselike sigmodontine rodents in South America (Hershkovitz 1966; Reig 1978; Marshall 1979; Webb and Marshall 1982). In the early part of the Cenozoic, South America was an island continent with a mammalian fauna composed primarily of marsupials, acquired before the breakup of Gondwana, the great southern continent. During the Cenozoic several lineages of mammals, including caviomorph rodents (guinea pigs and agoutis), primates, and xenarthrans (anteaters, armadillos, and sloths) colonized across the ocean. The isolation of South America ended about 3 million years ago, with the completion of the inter-American land bridge, and there was an exchange of mammals with North America. Even though approximately equal numbers of lineages moved from each continent into the other (Simpson 1980; Marshall et al. 1982), North American mammals were differentially successful in South America because of their higher speciation rates.

The sigmodontine rodents (cotton rats and rice rats) were among the North American lineages that moved south, and they probably colonized by dispersing over water even before there was a complete land bridge. The Sigmodontinae is now the most speciose group of land mammals in South America. The original colonists have speciated within the continent to produce at least 52 genera and more than 250 species (Hershkovitz 1966; Reig 1978, 1989; Marshall 1979; Webb and Marshall 1982). The exact mechanisms of this speciation are not known, but these rodents share with other members of the mouse family, Muridae, several traits that promote the generation and survival of new species. These include both attributes that limit dispersal and promote ecological specialization (e.g., in habitat and diet), and characteristics that make them excellent colonizers and able to maintain viable populations in small, isolated patches of habitat. Whatever the exact mechanisms, the fact that sigmodontine rodents and a few other lineages of recent colonists have speciated at much higher rates than the endemic lineages has resulted in the gradual replacement during the last 3 million years of the original South American mammalian fauna by one dominated by groups of North American origin.

One of the triumphs of modern biology has been the development of theoretical and molecular methods to reconstruct phylogenetic history (e.g., Hennig 1966; Wiley 1981; de Queiroz and Gauthier 1992; Huelsenbeck and Hillis 1993). Once phylogenies have been derived, it is possible to apply analytical methods to identify the origin and follow the evolution of particular traits or suites of characteristics. Thus, Doyle and Donoghue

(1986, 1987; Doyle 1978; Donoghue 1989) have not only clarified the early evolutionary history of the angiosperms, but have also identified combinations of reproductive traits and other characteristics that may have contributed to the spectacular success of certain lineages. A similar approach is the historical and phylogenetic comparative animal biology being developed by Brooks and McLennan (1991), Harvey and Pagel (1991), and others.

The prospects are exciting, especially when the insights of comparative phylogenetic studies are combined with those from the fossil record and from contemporary ecology and biogeography. We stand to learn a great deal not only about the facts of ancestor-descendant relationships and the patterns of diversity over time, but also about the ecological and evolutionary processes responsible for these changes.

SPECULATIONS ON THE EVOLUTION OF ENERGY USE, COMPLEXITY, AND SIZE

Evolutionary biologists, both micro and macro, have almost universally denied that there is any goal of organic evolution. Consequently, they have been very uncomfortable with the notion that there are consistent (i.e., orthogenetic) trends in evolution. While I do not want to take the extreme view and suggest that evolution is directed toward any particular goal, I think there are clear trends toward increasing energy use, complexity, and body size.

The energetic and thermodynamic perspective that I have begun to develop in chapters 5 and 10 helps to explain these trends. Living things are ultimately limited by their ability to acquire and transform energy to do the work required for growth, survival, and reproduction. The supply of usable energy is limited by the input of solar energy to the earth and by fundamental evolutionary and functional constraints on the mechanisms (primarily photosynthesis and respiration) by which organisms obtain and transform this energy.[4] To a large extent, the game of life played by all organisms is a competitive contest to obtain a share of this productivity, and relative success is measured by the size of the share (see discussions of reproductive power and the maximum power principle in chaps. 5 and 10). Those individuals within species and those species within biotas that obtain a larger share of the energy are, on average, able to produce more descendants. Because of the constraints on energy availability, most of the

---

4. A small fraction of the energy used by organisms, such as those inhabiting hydrothermal vents in the oceans, comes from chemosynthesis rather than photosynthesis, but this source too is limited by supply rates and biological constraints.

success of any individual or species will be achieved at the expense of other units at the same level of organization, causing the latter to obtain a smaller share of energy and to mobilize less reproductive power, thus reducing their fitness.

Further, increasing energy use and reproductive power seems to be accomplished by increasing complexity and body size. I submit that it is the large, complex units of biological organization, such as trees, large mammals, ant and termite colonies, and human society, that currently dominate energy use and most influence abiotic and biotic ecosystem processes. The largest and most complex systems maintain themselves farthest from thermodynamic equilibrium. They do so by taking in and transforming large quantities of energy. The power that is mobilized is used to enhance their own fitness. The work that is performed also transforms the larger system (population, community, ecosystem) in which they are embedded (Brown, in press).

Three additional comments are relevant here. First, these arguments seem to offer a way to reconcile the reproductive power model predicting an intermediate optimal body size for a species (chap. 5) with the empirical finding that species of large size dominate energy use in ecosystems. The model assumes that the species occurs in isolation. In competitive environments, where other species are present, large organisms are able to attain at least as much reproductive power as small ones. Second, I am aware that many small, relatively simple organisms with low rates of energy use have managed to survive for billions of years and to coexist with other organisms in contemporary ecosystems. I suggest, however, that they make their living either by utilizing energy sources that the large, complex, energetically powerful organisms cannot exploit because of evolutionary and functional constraints (e.g., archaebacteria and cyanobacteria in hot springs) or energy sources that are actually made available by larger and more complex organisms (e.g., microbial pathogens and detritivores). Third, the trend toward increasing energy use, size, and complexity does not imply that all systems that increase these traits will persist. The process of selection, whether it occurs at the level of individuals or species, does not include foresight. Systems that are highly fit at one level can fail to survive, either because of chance events or because of opposing selection at another level.

Thus I suggest that it is no accident that the largest and most complex organisms and societies that ever existed are present on the earth today. The 4-billion-year history of life reflects a relatively uninterrupted trend toward increasing energy use, size, and complexity by all living things collectively and by the individuals, societies, and species that make up the global biota. I suspect that these ideas will be controversial. Additional theoretical and empirical research must be done to make them rigorous and convincing. But I can see no other way to explain the apparent macroecological and macroevolutionary patterns.

CONCLUDING REMARKS

I have been able to cite only a selected few of the relevant studies that have contributed to the recent advances in our knowledge of the history of the earth and its organisms. And all signs suggest that this progress will continue, probably at an accelerating rate, for at least another decade or two. New data from paleontological and paleoenvironmental studies, new techniques for reconstructing phylogenies and for interpreting fossils, and the development and application of new hypotheses and concepts are all playing major roles in these accomplishments. The new breed of paleobiologists, macroevolutionists, and systematists is not content simply to describe the history of biological diversity, but increasingly seeks to develop and test mechanistic hypotheses to account for the changes.

Macroecology can play a major role in this endeavor. On the one hand, it can contribute to our understanding of the processes of organism-environment interaction that caused many of the changes that are recorded in the fossil record and in the phylogenetic relationships among species. On the other hand, it can contribute a historical perspective to our understanding of contemporary patterns of abundance, distribution, and diversity of species. Because of its emphasis on large spatial and temporal scales, on species as the units of study, and on kinds of data that can be relatively easily obtained—even from fossils—the macroecological research program can make a unique contribution to our understanding of the relationships between the present and the past.

12

Applications

Human Ecology And Conservation Biology

THE PROBLEM

The Ecological Dominance of Homo sapiens

Applying the approach and results of macroecology to practical problems requires that our own species study its impact on the earth and on the other species that share the planet with us. On the one hand, *Homo sapiens* is just another of the millions of species that inhabit the earth. *Homo sapiens* was formed by the same processes that produced all other species, and our abundance and distribution are governed by the same natural laws that affect all living things.

On the other hand, humans are different in at least two respects. First, ours is the only species in the history of life ever to have acquired such power to change the earth and such dominance over all other organisms. Second, ours is the only species with the intellectual capacity and technological tools not only to study itself and its world, but also to apply the resulting knowledge to change the world. The question is how, at this critical stage in the earth's history, we will use this power and knowledge.

The success of *Homo sapiens* is staggering. The species now numbers more than 5.5 billion individuals. It has a wider geographic distribution than any other organism, with established populations on all the continents and large islands and with dispersing individuals on and under the oceans, in the atmosphere, and even in space. No part of the earth has escaped human influence. Human impact is greatest on the approximately 12% of the earth's land area that has been converted to cities, suburbs, and intensive agriculture, but it has also altered a much larger area that is used for grazing livestock, harvesting timber and wildlife, disposal of toxic wastes, and other activities.

The rise to dominance of *Homo sapiens* has occurred in just a few tens of thousands of years. This dominance can be attributed largely to three technological inventions. The first was the use of tools and fire for hunting and gathering. These enabled aboriginal humans to manipulate vast landscapes and to trigger the extinction of many species, including the Pleistocene megafauna of several continents and the native animals on many islands (see below). The second technological breakthrough was the development of agriculture and the concomitant domestication of plants and animals. This enabled highly productive habitats to be used entirely to support dense human populations and led to the development of complex societies. The final innovation was the technology for harnessing the energy in fossil fuels and using it to build even more complex social, political, and economic systems based on industry and information. This enabled humans to attain their present precarious state of knowledge and power.

Human Ecology

Western societies do not have a well-developed tradition for studying the place of humans in nature. The natural sciences, in their search for rigor and objectivity, have tended to view everything influenced by humans as "artificial" and everything else as "natural." Nowhere is this more apparent than in the recent tradition of ecology. Most ecologists study wild organisms in "natural areas" where human influence is minimal. The study of humans and their interrelationships with the rest of the natural world has been left to the "social" and the "applied" sciences, both of which have been viewed with disdain by many of those who practice "pure" ecology.

In reality, of course, these are false distinctions. But we have not yet completely come to terms with the Darwinian revolution, its rejection of special creation, and its explicit recognition that *Homo sapiens* is just another species and an integral part of nature. Ecologists have studied the effects of starfish, largemouth bass, sea otters, beavers, and other "keystone" species that control the structure and function of ecosystems, but they have been strangely reluctant to study the most key species of all, their own. We know a great deal about the ecology of wilderness areas and natural reserves, but we are just beginning to study agroecosystems. Even more serious, given that more than 75% of the human population lives in or immediately adjacent to towns and cities, is the almost total lack of studies of urban and suburban ecosystems (McDonnell and Pickett 1990). Society can no longer afford the luxury of "basic" ecology that continues to avoid studying the human species, its ecological niche, and its effects on other organisms and the abiotic environment (see Lubchenco et al. 1991).

One thing that I advocate, then, is an increased emphasis on human ecology. Ecologists—and biogeographers and evolutionary biologists as well—should begin to study humans and the entire spectrum of human-influenced ecosystems with the same zeal and objectivity that they have applied to the study of wild species and relatively pristine habitats. Nearly two decades ago Eugene Odum (1969) called for a similar research program, but there was little response. These ideas should get a more favorable reception now.

Many of the needed approaches to human ecology will not be macroecological, however. At least they will not be the kind of macroecology discussed in this book. They will be basic studies of the niches and the environmental impacts of different populations of *Homo sapiens* in different places. They will be applied studies on how to manage human social, political, economic, and technological activities and how to mitigate the deleterious effects of these activities on our own species, other organisms, and the abiotic environment.

Nevertheless, the macroecological perspective developed in this book does have much to contribute. Macroecology has had to confront four issues that will also face efforts to study human ecology. First, controlled, manipulative experiments will often be impractical or immoral, so other ways of detecting patterns and testing hypotheses will have to be developed. Second, the statistical search for emergent patterns and processes should often offer an alternative to experimental manipulation—and fortunately there are lots of statistics on humans and their activities. Third, the most serious effects of human activities occur on regional to global spatial scales and on at least decadal temporal scales—much larger scales than most traditional ecologists have studied. And finally, a thorough study of human ecology will require a broad, interdisciplinary approach, involving both natural and social scientists: ecologists, evolutionary biologists, behaviorists, physicians, paleobiologists, earth scientists, geographers, psychologists, sociologists, economists, and political scientists. The macroecological research program can also be applied in much more direct ways, however, to provide a "big picture" perspective on the effects of humans on other organisms.

THE LOSS OF BIOLOGICAL DIVERSITY: CAUSES

A Historical Perspective: Extinctions Caused by Aboriginal Cultures

Many individuals have recently called attention to the accelerating wave of human-caused extinctions that is depleting the biological diversity of the earth (e.g., Diamond 1984; Myers 1987; May 1988; Wilson 1988). It is important to realize, however, that *Homo sapiens* has been causing extinctions for a long time. Probably ever since humans evolved, and certainly ever since they developed tools and fire, they have been increasing their population, expanding their geographic range, using an increasingly wide range of resources, altering habitats, and influencing other organisms by hunting and gathering, competition for resources, and other activities. The result has been a wave of human-caused extinctions that began at least tens of thousands of years ago and has built up to its present magnitude.

Perhaps the earliest human-caused mass extinctions occurred in the Pleistocene. Humans played a major role in the extinction of the Pleistocene megafauna (the giant mammals and flightless birds) of North and South America, northern Eurasia, perhaps Australia, and many islands, from Madagascar and New Zealand to tiny Pacific atolls.

The North American example is illustrative. Sometime within the last 30,000 years *Homo sapiens* colonized North America via the Bering land bridge that connected Siberia and Alaska. Approximately coincident with the spread of humans across the continent were the extinctions of many large mammals, including mammoths, mastodons, gomphotheres, giant ground sloths, horses, camels, giant bison, sabretooth cats, and dire wolves (Gilbert and Martin 1984; Martin 1984a,b, 1990). There can be no doubt that humans successfully hunted many of these large mammals. A substantial fraction of fossil mammoths that died within the last 15,000 years are human kills, and there is evidence of human predation on several other species of large, now extinct herbivores (Martin and Klein 1984). On the basis of this and other evidence, Martin (1966, 1967) advanced the overkill hypothesis: that severe predation on naive animals along the front of human colonization directly caused the extinction of a large proportion of the Pleistocene fauna.

There are problems with this extreme version of the human predation hypothesis. Several authors (see discussion by Guilday 1984; Grayson 1984; Stuart 1991) have attributed the late Pleistocene extinctions primarily to climate change. They cite the following evidence: (1) many of the large mammals that went extinct were species of open grassland, savanna, and steppe habitats; (2) global warming signaling the end of the Pleistocene began about 12,000 years ago; and (3) at about that time there was an expansion of birch, spruce, and other forests and a contraction of open habitats.

Further, Grayson (1977, 1984) calls attention to the disappearance of many bird species at about the same time as the extinction of the giant mammals. These birds included a stork, hawks, eagles, vultures, teratorns (giant scavengers), a lapwing, a thick-knee, jays, and blackbirds (Steadman and Martin 1984). While many of these were not the kinds of birds likely to have been hunted to any great extent, several were almost certainly carrion feeders whose demise may have been triggered by a decline in food supply caused by the megafaunal extinctions. Many of the remaining species were probably found in open grassland or shrub-steppe habitats.

While the changes in vegetation and habitat are becoming increasingly well documented, the evidence that they were caused entirely or even primarily by climate change is less convincing. Indeed, one of the striking features of Pleistocene megafaunal extinctions throughout the world is that they were not contemporaneous on different continents and large islands and did not appear to be coincident with a consistent climatic signal, but in many places were closely associated with human colonization.[1]

How, then, do we account for the megafaunal extinctions in North America and perhaps elsewhere? I find a recent hypothesis advanced by Owen-Smith (1987, 1988, 1989) attractive. Owen-Smith studies the ecology of surviving "megaherbivores"—elephants, rhinoceroses, elands, and so forth—in Africa. He suggests the following scenario for Pleistocene North America. Hunting by colonizing humans reduced or eliminated the largest mammalian herbivores. The browsing and disturbance of these species had played a major role, as it does in parts of Africa today, in suppressing trees and maintaining the habitat as a grassland or steppe. The reduction in the "keystone" megaherbivores, perhaps in conjunction with climate change, caused an expansion of forests and a shrinkage of open habitats. These habitat changes caused populations of many species of both mammals and birds to become reduced and fragmented, and this, together with additional human hunting, loss of prey and carrion for predators and scavengers, and other changes, eventually led to the extinction of additional species. Thus, Owen-Smith suggests that human colonization of North America was the trigger that set in motion a cascade of events that eventually caused the extinction of the megafauna and many other species in the late Pleistocene.

1. Two other points are relevant here. First, aboriginal humans colonized the Australian subcontinent, and many species of giant marsupials went extinct there, during the late Pleistocene. The role that humans may have played in these extinctions is debatable, but both events occurred well before the arrival of humans and the disappearance of the megafauna in North America. Second, no mass extinction of Pleistocene megafauna apparently occurred in Africa (at least, a diverse assemblage of giant mammals has survived). Martin, Diamond, and others have hypothesized that this is because humans originated in Africa. Having the opportunity to coevolve with *Homo sapiens,* many African species were presumably able to adjust to the relatively gradual changes in habitats and hunting methods. In contrast, when humans expanded their geographic range onto new continents and islands, many species could not cope with the large, rapid changes in their environment that resulted.

If this scenario of *Homo sapiens* as a keystone species in Pleistocene North America is controversial—and it is—there can be no doubt that colonizing humans hunted birds and mammals to extinction on many islands. Not only did humans kill off the giant flightless elephant birds and moas on Madagascar and New Zealand, respectively, they caused the extinction of even small passerines on small Pacific islands (Olson 1989; Steadman 1989). The work of S. Olson and H. James in the Hawaiian Islands provides perhaps the best-documented example. The first wave of extinctions began about a thousand years ago with the arrival of the Polynesians and the beginning of hunting and deforestation (Olson and James 1982a,b, 1984). Conspicuous among the species that disappeared were some very large, sometimes flightless birds as well as small passerines. A second wave of extinctions followed European settlement, caused primarily by a combination of habitat destruction and introductions of mammalian predators (mongooses, rats, and feral cats), avian competitors (a host of alien passerine bird species), and at least one parasite (avian malaria). The result has been that out of an original land bird fauna of at least 99 native species, 50 were extirpated by the Polynesians, 17 have disappeared in the approximately 100 years since European settlement, and 19 more are currently in grave danger of extinction. Of the original species of native Hawaiian birds, 87% have either gone extinct or become endangered as a result of human activities!

One important feature of the mammal and bird extinctions attributed to aboriginal humans is that they have been highly selective in several respects. First, mammals and flightless birds of large body size have been differentially—sometimes completely—eliminated from all continents, except for Africa,[2] and from many islands (Diamond 1984; Martin and Klein 1984; Owen-Smith 1988). Furthermore, most surviving species of giant mammalian herbivores (elephants and rhinoceroses) and many other species of large mammals and birds are currently endangered. Second, endemic species inhabiting islands have been differentially eliminated. Although we are just beginning to appreciate the extinctions of island species caused by aboriginal humans, there can be no doubt that European colonists havé caused an additional wave of extinctions that continues right up to the present. Third, species with certain ecological affinities and requirements were differentially lost from the fauna. For example, the late Pleistocene extinctions in North America selectively eliminated both bird and mammal species of grassland and steppe habitats.

2. See note 1 above.

Current Extinctions and Human Resource Use

From these ancient beginnings, the wave of human-caused extinctions has continued to build. Modern humans, increasing exponentially in population, still colonizing new areas, and now using the marvels of contemporary technology, are having an even more devastating effect on the earth's biota than their ancestors did. The rate and absolute magnitude of contemporary extinctions are approached only by the few, infrequent mass extinctions recorded in the fossil record.

The proximate causes of these extinctions are many: hunting and gathering, habitat destruction and fragmentation, pollution, and the effects of domestic and exotic species. There is just one ultimate cause, however: the growth and increasing impact of the human population. A macroecological perspective immediately makes it clear not only that the present magnitude of extinction is inevitable in this situation, but also that many more species are doomed to extinction if present trends continue.

One simple but powerful argument has been made by my former student David Wright (1987, 1990). It is based on the fact that humans have converted about 12% of the earth's most productive habitats to agriculture, suburbs, and cities (see also Vitousek et al. 1986). Wright uses species-area and what he calls species-energy relationships (Wright 1983) to calculate the reduction in species richness predicted from such changes in land use—from supporting diverse assemblages of plants, animals, and microbes to supporting simplified, low-diversity systems to sustain human populations (table 12.1). Taking a similar approach, Vitousek et al. (1986) and Wright (1990) estimate the total fraction of the earth's terrestrial productivity that has been preempted by *Homo sapiens* at between 20% and 30%—perhaps as high as 39%!

E. H. Rapoport (pers. comm.) and Wright (1990; see table 12.2) have estimated the populations and energy use of humans and their most common domesticated animals. On this basis, *Homo sapiens,* cattle, and other

TABLE 12.1. Global species richness of land mammals, land birds, butterflies, and angiosperm plants and estimated loss of species predicted from human effects on land use (area) and productivity (energy).

Taxon	Recent richness	Estimated loss of species based on	
		Area	Energy
Land mammals	3,094	168 (5.4%)	310 (9.7%)
Land birds	7,860	527 (6.7%)	910 (11.6%)
Butterflies	13,000	1,250 (9.6%)	9,000 (15.3%)
Angiosperms	250,000	14,500 (5.8%)	24,100 (9.6%)

Source: Wright 1987.

TABLE 12.2. Total world population, per individual energy consumption, total global energy use by humans and domestic livestock.

Species	Total population (10^9)	Energy use per individual (10^9 J/yr)	Total energy use (10^{18} J/yr)
Humans	5.50	4	22.0
Cattle	1.27	47	59.7
Buffalo	0.14	68	9.5
Sheep	1.15	7.9	9.1
Pigs	0.82	10	8.2
Chickens	9.35	0.62	5.8
Goats	0.49	6.1	3.0
Horses	0.07	41	2.7
Other livestock	0.07	40	3.0

Source: Data from Wright 1990.

livestock are the earth's dominant animals. A similar comparison for plants would certainly show at least comparable global dominance by the cereal grains (rice, wheat, corn, millet, sorghum, and oats) and a few other species, such as sugarcane and potatoes. Human activities have also benefited many species, such as commensals (e.g., house mice and houseflies), pets, ornamental plants, and agricultural weeds and insect pests, that are not used primarily for food. For example, domestic dogs and cats are by far the most abundant and geographically widespread mammalian carnivores.

Wright's predictions are based only on the conversion of land and productivity to human-supporting agro-urban ecosystems. His analysis does not include the extinctions caused by hunting and fishing, forestry, wood-cutting, pastoral grazing, desertification, human-assisted spread of parasites and diseases, and introductions of exotic species. Nevertheless, the results are sobering (table 12.1). Between 9% and 15% of the vascular plant, butterfly, bird, and mammal species are predicted to go extinct. This is only a fraction of the extinctions of species in these groups that have actually occurred in the last few centuries. Therefore, Wright suggests that many species are "zombies," or living dead, that are doomed but have not yet experienced the drought, disease, or stochastic fluctuation that will push them to their inevitable extinction.

Note that Wright's argument is macroecological in that it is based on the statistical distribution of resources and geographic space among species. It emphatically makes the point that our own species has monopolized the productivity of the earth, changing the distribution of resources among the other species and dooming many of them to extinction. There is simply no way that the earth can continue to support both the present population of *Homo sapiens* and the present diversity of other species. This point, which comes from a macroecological perspective, has not yet been appreciated by many well-meaning environmentalists and conservation biologists. It is why most conservation efforts will ultimately fail unless the human population and its demands for resources can be reduced.

MACROECOLOGICAL PERSPECTIVES ON CONSERVATION

While we are waiting, hoping, and trying to bring human activities into a sustainable balance with the earth's environment, less ambitious conservation activities can play an important role in reducing the rates at which species are being lost and habitats are being degraded. There are several ways that a macroecological approach can help, both to define the problems and to seek practical solutions.

Assessing Vulnerability to Extinction

One clear message that comes from macroecological patterns (chaps. 1, 6, and 9), the fossil record (chaps. 9 and 11), and human history (this chapter) is that extinction is predictable. Certainly, extinction is an inherently probabilistic process with an element of apparent stochasticity. But the extinctions that occurred in the past, especially during periods of major environmental change, were quite selective: species with certain characteristics tended to go extinct, while other species, usually with opposite characteristics, tended to survive.

If those species most likely to be vulnerable to environmental change can be identified, then they can be singled out for special attention in the form of monitoring and management. A macroecological perspective identifies three characteristics of species that are associated with high probabilities of extinction: (1) large body size, (2) small geographic range and associated ecological specialization, and (3) insular endemism. This macroecological perspective is valuable for three reasons. First, it identifies characteristics that might be missed by a more microscopic approach to risk assessment. Much of the theory and data on extinction stresses the consequences of small population size and demographic stochasticity (e.g., Soulé 1980; 1986; Soulé and Wilcox 1980; Pimm 1991). While these are important proximate factors, they fail to take account of an important reality: many of the species that have gone extinct or become endangered within the last two centuries were originally extremely abundant and in no apparent danger. Examples from North America include the American burying beetle, Colorado squawfish, lake and cutthroat trout, several salmon species, Carolina parakeet, passenger pigeon, whooping crane, lesser prairie chicken, bison, sea otter, and northern elephant seal. While not all of these fall into one of the above three categories of vulnerability based on macroecological criteria, most of them do. In particular, many of them are of conspicuously larger body size than closely related species that have not experienced such drastic declines.

Second, these macroecological correlates of vulnerability emphasize the importance of considering several variables in assessing risk of extinction. If any single characteristic is used, many vulnerable species will be missed. For example, focusing on species with small geographic ranges and narrow habitat requirements would miss a large class, noted above, of extremely abundant, geographically widespread, and seemingly generalized species that have declined precipitously. An excellent example is the American burying beetle (*Nicrophorus americanus*), which once ranged over the entire eastern half of the United States and was very abundant until the early 1900s. Now it survives in small populations on the edges of its former geographic range: on Block Island, Rhode Island, and on prairie sites in Oklahoma, Arkansas, and perhaps Nebraska (Lomolino et al., in press).

Third, macroecological correlates of vulnerability can be used to make preliminary assessments of priority for monitoring and potential management when more detailed ecological data are not available. There is simply not enough time, skilled personnel, and money to make detailed field studies of the population biology of all species before they go extinct. Macroecological analyses of data that are easier and less costly to obtain offer a practical alternative. This is especially true for groups of organisms, such as most insects and plants, and for geographic regions, such as most developing countries, for which the biology of species is poorly known. Arita, Robinson, and Redford (1990; see also Dobson and Yu, in press) used essentially this approach in assessing the conservation status of a large number of South American mammal species (table 12.3; for an alternative approach see McDonald and Brown 1992 and chapter 1).

Finally, it is important to add one caveat. I do not think that the three macroecological correlates of extinction vulnerability identified above are adequately diagnostic. It should be possible to extend this macroecological approach, however, to find additional characteristics that can be used. I am optimistic that there are a relatively small number of species characteristics that can be easily measured and combined in a multivariate analysis to predict risk of extinction. This would give conservation biologists a valuable tool for making policy and management decisions.

Hot Spots and Biological Reserves

The extinctions caused or threatened by modern humans usually occur as a consequence of incremental range contraction and decrease in abun-

TABLE 12.3. The categorization of 100 species of tropical forest mammals from South America with respect to abundance, distribution, and body size.

Population density	Geographic distribution	
	Restricted	Widespread
High		
Number of species	29	21
Mean body mass	437 g	801 g
Low		
Number of species	21	29
Mean body mass	2,746 g	6,229 g

Source: Compiled from data in Arita, Robinson, and Redford (1990; see also Dobson and Yu, in press).

Note: Species were divided into four cells based on dichotomous categories of population density and area of geographic range, and the mean body mass of the species in each cell was calculated. Note that average body size increased with decreasing density and increasing range. Compare with the somewhat similar classification of British plants in table 4.1 and Rabinowitz, Cairns, and Dillon (1986).

dance. Risk of extinction can be reduced and costly efforts to save endangered species can be avoided if action is taken well before the situation has become critical. To do this requires an understanding of the spatial and temporal patterns of distribution and abundance and of the processes that produce these patterns. Since these patterns and processes largely reflect the extent to which the human-modified environment continues to meet the niche requirements of the species, the detailed information needed will always be species-specific—just like the recovery plan for an endangered species.

At another level, however, the macroecological perspective can provide several valuable insights. For example, the data presented in chapter 4 suggest that the spatial distribution of abundance of nearly all species is extremely heterogeneous. The orders of magnitude variation in abundance has important implications for conservation. Since only a small proportion of the sites where a species occurs contain a large proportion of all the individuals, it is important to identify these "hot spots." Then conservation reserves and ecosystem management practices can be designed so as to preserve the hot spots of as many species as possible, with priority assigned to species considered to be especially vulnerable to extinction.

There should be some relatively low-cost ways of doing this. First, because of the magnitude of the variation, hot spots can be identified using relatively imprecise census information. Second, the wide variation in abundance can be used to guide the search for favorable environmental conditions. I can best illustrate with an example. Dave Mehlman, George

Stevens, and I are using BBS data to characterize the spatial distribution of abundances of breeding birds (e.g., fig. 4.6). We then plan to obtain data on environmental variables for the BBS census routes, and to use multivariate statistical procedures to derive sets of variables that are correlated with the abundance of each species. Thus, we can begin to characterize the niches of the species. Initially this will be a "brute-force" approach based entirely on correlation. But with additional work, we should be able to identify the environmental variables that actually limit abundance and distribution.

But even this initial correlative approach should have valuable conservation applications. It should be adequate to identify important environmental characteristics of hot spots and to search for additional hot spots (especially for species of conservation concern) in areas that have not yet been surveyed. A relatively good characterization should be possible by using the data bases on climate, topography, geology, soils, vegetation, and other environmental features that are becoming available from both ground-based sources and remote sensing. Geographic Information Systems, geostatistics and multivariate statistics, and other computer technology can be used to analyze the data. In this way it should be possible to obtain answers to several important questions: How many hot spots for how many different species are included in existing or proposed reserves? To what extent are the hot spots of different species coincident? What environmental conditions are correlated with the hot spots, especially those of species of particular concern?

It is important to emphasize that this macroecological approach takes into account both the uniqueness of species and their niches and the emergent statistical distribution of abundance that is apparently characteristic of all species. While the location and environmental characteristics of hot spots may turn out to be highly individualistic attributes of species, the same data sets, technological tools, and statistical methods can be applied to a large number of species: all those whose abundances and distributions are reasonably well surveyed. Thus, this one macroscopic approach potentially offers a relatively fast and inexpensive way to address conservation problems faced by many species.

The wide spatial variation in the abundance of most species suggests a possible concern about conservation plans based solely or largely on considerations of species diversity: Do the sites of high diversity being targeted for reserves and management activities also support high abundances of the species? I am afraid that the answer will often be no, because sites of exceptionally high diversity will often contain a mosaic of small patches of many different kinds of habitat or be near the edges of the ecological and/or geographic ranges of many species. Note, however, that this is an empirical question, and one that should not be difficult to answer using macroecological approaches.

The answer will be important, because it would be counterproductive to spend valuable resources to preserve areas that support such small populations of most species that many extinctions are likely. Increasingly government agencies, academic scientists, and nongovernmental conservation organizations are seeking robust, practical, low-cost methods of obtaining data and using them to set management and policy priorities. Most of these are necessarily macroscopic approaches. An example is gap analysis (e.g., Scott et al. 1993), an approach that uses the potential presence or absence of species (often compiled from maps of geographic ranges) in conjunction with measurements of environmental variables to locate sites of high species diversity for potential reserves and management activities. Gap analysis is inherently macroecological and has much in common with the approach of identifying and characterizing hot spots on a species-by-species basis. It differs, however, in its reliance on overall species diversity rather than on patterns of abundance within species. I think that this is a reason for concern, and I caution against uncritical application of gap analysis and other diversity-based approaches until their premises have been evaluated.

Patterns of Range Contraction

Macroecological approaches can also provide potentially valuable insights into the patterns of geographic range contraction that often occur as a result of human activities. Is there any general pattern? There are many examples of species contracting to sites near the centers of their ranges, perhaps to some of the hot spots where they were formerly most abundant. There are also examples, however, (e.g., the American burying beetle mentioned above, and New Zealand birds and reptiles that survive only on tiny offshore islands) of species contracting to sites at the edges of their original geographic ranges.

Both of these cases can be interpreted in terms of the empirical patterns of abundance and the niche-based theory presented in chapter 4. When there is a general degradation in the environment, and especially when human activities have affected several different niche variables, the effect should be to reduce the suitability of most sites. But hot spots should on average remain relatively more favorable, and the range should contract from the edges toward the center—or at least toward regions where hot spots are concentrated.

On the other hand, when there are drastic changes in just one or a small number of critical variables, then I expect exceptions to this pattern.

These conditions seem to fit most of the examples. For example, the survival of some bird and reptile species on tiny islands off the coast of New Zealand can be attributed to the establishment elsewhere throughout their former ranges of exotic predators, such as rats, cats, and possums (e.g., Atkinson and Bell 1973; Diamond and Veitch 1981). Similarly, it seems straightforward to predict that if severe global warming occurs, many species should contract from the lower latitudinal and elevational edges of their ranges—and also perhaps expand their ranges along the opposite borders (see also chapter 1 and Peters and Lovejoy 1992; Kareiva, Kingsolver, and Huey 1992; Mooney et al. 1993).

Exotic Species and the Cosmopolitization of the Earth's Biota

As the example of New Zealand birds illustrates (see also Drake et al. 1989; Hengeveld 1989), many conservation problems have been caused by colonizing exotic species. The most devastating invader usually has been *Homo sapiens,* but the effect of our own species has been compounded by our intentionally or inadvertently transporting many other organisms across biogeographic barriers. There has been an almost universal tendency to regard these invaders as "bad" and to place a high priority on the eradication of exotics and the preservation of native species. I don't want to argue against this position, but I do want to temper it with three bits of macroecological perspective.

The first point is that some substantial degree of "cosmopolitization" of the earth's biota is probably inevitable. Given their magnitude, human movement and commerce are almost certain to result in increased dispersal rates and in the eventual establishment of some proportion of the colonizing species. Furthermore, the habitat changes caused by human activities tend to favor establishment of exotic species, especially those with an evolutionary history of close association with humans, often at the expense of native species that have had less opportunity to adapt to human influences.

In chapter 9, I showed how the history of fragmentation of regions and habitats has played a major role in the creation and maintenance of global biological diversity. Isolation and speciation have resulted in the buildup of endemic biotas. Human-aided transport of species must have the opposite effect by crossing the historical biogeographic barriers to dispersal, reducing the biologically effective isolation of regions, and ultimately reducing global diversity by replacing some proportion of the endemic native species with invading exotics. In the absence of major human-caused disturbance, it might seem that the native species inhabiting the isolated patches would be able to resist invasion, because they have had ample opportunity to adapt to one another and to the local abiotic environment.

But experience with introduced species suggests just the opposite.[3] Frequently alien species are not only able to invade, but their colonization leads directly or indirectly to the extinction of native species. Oceanic islands are the most frequently cited examples of isolated habitats whose endemic species are being extirpated and replaced by invading exotics (Carlquist 1965; Williamson 1981).

However, freshwater fishes provide perhaps even more dramatic examples. A common set of fish species, including rainbow and brown trout, largemouth bass, bluegill, carp, tilapia, and mosquitofish, has been introduced into lakes and streams throughout the world to provide sport fisheries or for the control of aquatic plants and insects. Other fishes have been introduced inadvertently by release of aquarium and bait fish and by the connection of previously isolated bodies of water. Although these exotics have sometimes been able to coexist with native species, often they have caused extinctions. One particularly graphic example is the extinction of three native species in the Crystal Pool, an isolated desert spring near Death Valley, caused by the introduction of a handful of largemouth black bass (Soltz and Naiman 1978; see also Zaret and Paine 1973). This is comparable to the role of the infamous lighthouse keeper's cat, which single-handedly eliminated an endemic wren species from Stephen Island off New Zealand (MacArthur 1972b).

The potential magnitude of the cosmopolitization of the earth's biota can be estimated by the following macroecological thought experiment. Imagine that continental drift could be reversed and all of the earth's land could be reassembled in a single giant continent.[4] What would be the effect on global species diversity? I can make a quantitative estimate, because there is a species-area relationship for the continents (fig. 12.1). By extrapolating to the landmass area of the megacontinent, we can predict the total global species diversity for several groups of terrestrial organisms and compare those values with the actual values (Wright 1987). The results (table 12.4) are impressive: from 35% to 70% of existing species would be lost. For anyone who doubts that such reductions in diversity would occur, imagine what would happen if the Hawaiian Islands were united with a continent. Even if the joining occurred at a place of similar climate and habitat—say, Costa Rica—nearly all of the insular endemics

3. The same pattern obtains in invasions that were not human-assisted. An excellent example is the differential replacement of native South American mammals by lineages of North American ancestry following the connection of the two continents by a land bridge about 3 million years ago; see chapter 11 and references therein.

4. Such a supercontinent, called Pangaea, actually existed, but about 250 million years ago it began to break up and the fragments drifted apart.

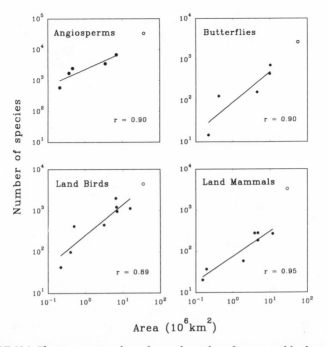

FIGURE 12.1. The species-area relationship, with number of species and landmass area plotted on logarithmic axes, for land mammals, land birds, butterflies, and angiosperms on continents and very large islands. The number of species in the world biota is plotted with an open circle and not included in the calculation of the regression line. Note that in all four groups, but especially in mammals and angiosperms, the actual number of species in the world is more than expected if the land areas were all joined in one supercontinent (see also text and table 12.4). (Adapted from Wright 1987; reprinted by permission of the Society for Conservation Biology and Blackwell Scientific Publications, Inc.)

would be rapidly replaced by continental species. After all, moving Hawaii to the continent would just accelerate the replacements of endemic species that have been occurring for the last century because of the transportation of continental species to Hawaii.

The second point is that all invading exotics should not automatically be considered to be bad. For one thing, while colonizations and expansions of geographic ranges must ultimately decrease global diversity, they tend to increase local diversity. The effect of high immigration rates in promoting local diversity is seen in the much higher alpha diversity on continents than in comparable habitats on small, isolated islands.[5] If local

5. The effect of immigration in increasing local diversity is also a direct prediction of MacArthur and Wilson's (1963, 1967; see chap. 9) equilibrium model of island biogeography. If the immigration rate is increased and everything else remains constant, the number of species at equilibrium is predicted to increase.

TABLE 12.4. The present number of species worldwide and the estimated reduction in species diversity in four groups of well-studied terrestrial organisms (mammals, birds, butterflies, and angiosperms) that would be predicted if all of the earth's land surface were joined in one supercontinent, but the present distribution of geological and climatic features were retained.

	Number of species		
Group	Present	Predicted	% of decrease
Land mammals	3,092	1,062	65.7
Land birds	7,860	4,120	47.6
Butterflies	13,000	8,447	35.0
Angiosperms	250,000	73,861	70.5

Source: Compiled from species-area relationships for the continents calculated by Wright 1987.

Note: This table gives a conservative estimate of the extent to which fragmentation and endemism contribute to diversity because it does not account for isolation of ranges within continents.

species diversity promotes ecological stability, then there might even be some long-term benefits of the cosmopolitization process.[6]

Exotic species can also play important roles in the maintenance and management of ecosystem processes. Because of the extinctions and permanent range contractions of many native species, there may often be a choice between exotics or no organisms at all to represent certain taxonomic groups and to perform important ecological functions. One of the best examples is the role of exotic fish species in the reservoir lakes of the southwestern United States. The Bureau of Reclamation and Army Corps of Engineers created these lakes by damming large rivers, such as the Colorado, Rio Grande, and their tributaries, to impound water for irrigation, hydroelectric power, and recreational opportunities. While an ecologist might question the wisdom of embarking on these projects, once the dams were in place, they produced a unique habitat—large bodies of deep, still water with a permanent thermocline—in a region where there had never been such lakes before, and where there were no native fishes adapted to such lacustrine environments. So entire fish communities were created by introducing many exotic species, including largemouth black bass to serve as the top predator and premier sport fish. Despite a seem-

6. Species diversity does appear to contribute to certain kinds of ecological stability. For example, as suggested by figure 12.2, habitats and regions with high species diversity appear to be more resistant to invading species, and they may also be less susceptible to disruptions of ecosystem processes caused by exotics. Further, the opposite effects of colonization on local and on global diversity make value judgments difficult. Like most conservationists, I would like to preserve as much global diversity as possible, but I think that conservationists must confront the reality of the cosmopolitization phenomenon. I do believe that there are much more serious threats to global biological diversity than the spread of exotic species.

ingly serendipitous history of introductions and other management decisions, these exotic communities work surprisingly well, playing key roles in limnological processes, providing food for a diverse assemblage of vertebrate predators, and supporting a multimillion-dollar sportfishing industry.

I could cite many other examples: the introduced birds that now make up virtually the entire avifauna of the lowlands of Hawaii (Moulton and Pimm 1986, 1987); the exotic species used for biological control of exotic pests that left their natural enemies behind when they colonized new regions (e.g., Dodd 1959; Harris, Peschken, and Milroy 1969; Drake et al. 1989; Hengeveld 1989); the domestic livestock used to "replace" extirpated native megaherbivores in order to maintain certain habitats in the southwestern United States or to restore tropical deciduous forests in Central America (e.g., Janzen 1982); the aesthetically pleasing, fast-growing, pollution-resistant "ornamental" plants that have been used to landscape city parks and private gardens throughout the world (Rapoport, Diaz Betancourt, and Lopez Moreno 1983); and the introduced grasses that have been used to reclaim strip-mined lands in arid regions (J. MacMahon, pers. comm.). Let me be clear at this point that I do not advocate wholesale importation of exotics. The literature of ecology is filled with examples of the damage caused by the unanticipated effects of deliberately introduced species. But since continuing extinctions of native species and expansions of exotic species seem to be inevitable, it is important to adopt a balanced view that recognizes the beneficial as well as the detrimental roles of exotic species.

The third point is that there is much to be learned by studying exotic species. It will be particularly valuable to investigate the characteristics of the species themselves and of their new environments that determine their success or failure in initial colonization, and—if they become established—that affect their rates of expansion, their new ecological and geographic ranges, and their effects on both other organisms and ecosystem processes. The resulting information can be used to assess the vulnerability of different geographic regions, habitat types, and native species to invading exotics. These assessments can then be used to set policies to reduce the chances that damaging exotics will be intentionally or accidentally introduced, and to detect and eradicate potentially damaging exotics before they become well established.

Elsewhere (Brown 1989), I have applied to exotics a recurrent theme of this book—that while many details are unique to individual species or particular environmental settings, there are also some emergent statistical

---

The invasion of exotics is more often a symptom of other kinds of human disturbance, disruption, and destruction than a serious cause of such changes.

patterns or rules. Thus, on the one hand, any effort to predict and/or limit the spread and impact of an invading pest must necessarily investigate the unique biology—and the unique ecological niche—of that species. If the potential damage is great, it may justify the kind of intensive, costly study that likely will be needed. On the other hand, biological invasions also show some general patterns that have practical implications. Compilations and analyses of data on exotic species from a macroscopic, statistical perspective can be very informative.

One example of such a pattern is the relationship, found by Keith Gido (pers. comm.) and shown in figure 12.2, between the number of established exotic fish species and the number of native species inhabiting a large watershed. While we would probably want to fit a constraint envelope rather than a regression line to this pattern, clearly drainages that have few native species are far more susceptible to invasion. Presumably this reflects the lower resistance of historically isolated, low-diversity biotas, such as those of the islands mentioned above. For another example, Brian Maurer and M. Villard (in press) have mapped the pattern of historical range expansion of the starling (*Sturnus vulgaris*) following its introduction into North America onto a map showing its present pattern of abundance in BBS surveys (fig. 12.3). The result suggests that in the early 1900s the starling expanded its range most rapidly into those regions that currently support the highest population densities. This pattern is consistent both with the niche-based view of distribution and abundance pre-

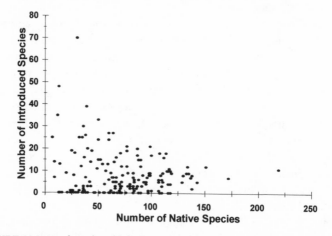

FIGURE 12.2. Number of established introduced species as a function of the number of native species of freshwater fishes in 135 large watersheds in temperate North America (not including streams in Florida). Note that the data points appear to fall into a constraint envelope: watersheds with few native species may vary greatly in their susceptibility to colonization by exotic species, whereas river systems with large numbers of native species are relatively resistant to invasion. (Courtesy of K. Gido.)

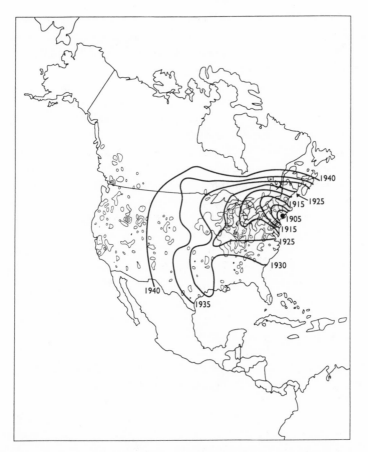

FIGURE 12.3. The pattern of expansion of the geographic range of the Eurasian starling (*Sturnus vulgaris*) following its introduction into North America at the turn of the century is superimposed on its current pattern of abundance (from Breeding Bird Survey data). Note that in eastern North America the starling appeared to colonize more rapidly in areas that presently support higher population densities. (Adapted from Maurer and Villard, in press.)

sented in chapter 4 and with the patterns of range contraction in declining species mentioned above.

Most immigrating exotics fail to become established (e.g., Moulton and Pimm 1986, 1987; Drake et al. 1989; Hengeveld 1989). It is becoming possible to identify variables that affect the probability of success. These include: (1) the number and composition of immigrants, with large absolute and effective population sizes more likely to succeed; (2) the environment of the region being invaded, with small, insular regions or habitats containing relatively few, highly differentiated, endemic species being more susceptible to invasion; (3) the source of the immigrants, with those from larger areas with similar abiotic conditions and from more diverse biotas being more likely to succeed; and (4) the characteristics of the invading species themselves, with "generalized" species that are abundant and widespread in their native regions being more likely to invade other regions. I present these as tentative, qualitative generalizations, but ones that can be framed and tested as hypotheses. For example, a special case of variable 4 above is the testable prediction that species possessing traits that reduce competition with already established species, both natives and exotics, should on average be more likely to colonize. Moulton and Pimm (1986, 1987; see also Simberloff and Boecklen 1991; Lockwood, Moulton, and Anderson 1993; Lockwood and Moulton 1994) have made, tested, and supported this prediction by analyzing the morphological characteristics of successful and unsuccessful avian colonists of Hawaii, Fiji, and Bermuda.

Information on successful and unsuccessful establishment of exotics should also have useful applications in deliberate efforts to introduce species: for example, to serve as agents of biological control or to reestablish populations of endangered species at sites within their former geographic ranges.

CONCLUDING REMARKS

In chapter 1, I began this book with an example of how the macroecological research program has addressed one specific conservation problem: how many and which of the mammal species inhabiting isolated desert mountains would be predicted to go extinct under one scenario of global climate change. This example illustrates one application of the large-scale, statistical, interdisciplinary approach of macroecology to a potentially important problem in conservation biology.

In this penultimate chapter, I extend this approach. I begin by placing the ecology of *Homo sapiens* and the impacts of modern humans on other organisms in a general, macroecological perspective. This treatment emphasizes the extent to which our species dominates use of the earth's resources. It also suggests that many of the impacts of humans, such as the destruction of habitat, extinctions or range contractions of native species, and invasions of exotic species, will continue to increase for at least the next few decades. Ultimately, the only way to preserve most of the earth's remaining biological diversity is to reverse human population growth and reduce human use of the earth's resources. In the meantime, application of macroecological concepts and data can help us to understand the changes that have occurred, to predict those that are likely to occur in the future, and to manage species and ecosystems so as to slow and resist some of the most damaging changes.

13   Reflections and Prospects

In closing this book, I want to comment briefly on what I believe are its major contributions. I also want to say a few words about the future of macroecological research.

WHAT IS NEW?

The Questions

Most of the questions posed in this book are not really new. As I pointed out in chapter 2, many of the questions that interest me were asked by Darwin and Wallace, investigated by Grinnell and Lotka, and posed in essentially their current form by MacArthur, Hutchinson, Preston, and Williams. I continually reread the writings of these predecessors to obtain new insights. An increasing number of my contemporary colleagues are addressing these questions in ways that certainly fit my definition of macroecology.

For me the ultimate question was framed by Hutchinson (1959) when, in "Homage to Santa Rosalia," he asked why there are so many kinds of organisms. Critics have complained that this is not really an answerable question. While they have a point, they miss the important one. The question of what produces and maintains the biological diversity of the earth is a big, complex, interdisciplinary question—and at present it is not possible to give a neat, concise, definitive answer. It is possible, however, to break this cosmic question down into a number of more specific questions that can be answered, and then—here is the most difficult part—to try to put these pieces together to begin to construct the big picture.

The questions are relatively straightforward. What are the intrinsic characteristics of species and the extrinsic characteristics of their environment that interact to regulate abundance over time and distribution over space? What are the relationships of species with one another and with their abiotic environment that determine the number and kinds of species that coexist in one place? How do these relationships cause changes in the number and characteristics of species, both in space, across the landscape and over the globe, and in time, over the history of life on earth and over much smaller time frames? How are new species produced, and what determines how they differ from their ancestors and whether they survive, spread, and give rise to descendant species?

I can claim only modest progress toward answering these questions, and equally modest progress in synthesizing the answers to address Hutchinson's big question.

Approach and Methodology

What I have done in this book is to develop a conceptual and operational framework for tackling some of the longstanding questions about the abundance, distribution, and diversity of species. Even here I make little claim to novelty. I have read extensively—but not as much as I would have liked—and borrowed shamelessly from both predecessors and con- temporaries. In doing so, I have developed ways of thinking about the questions, searching for patterns in data, and framing and testing hypoth- eses. Some of these are new to me, and they are responsible for the feel- ing of excitement and progress that I have tried to convey in the preced- ing chapters. What are they?

One is the focus on emergent properties of large numbers of individu- als or species as reflected in the statistical distributions of variables. This is hardly new as a practice in ecological research, but I have made it a central theme of macroecology. In doing so, I have tried to develop a conceptual rationale for this approach as one way to search for the laws that govern the structure and dynamics of complex ecological systems. Recurrent patterns of distributions of body sizes, abundances, and geo- graphic range sizes among species have intrigued ecologists for decades. Further work only more strongly suggests that these patterns are among the most general and important features of the organization of the living world.

Mechanistic explanations for these patterns, however, have remained elusive. Given their pervasive generality, it is surprising that so few inves- tigators have made sustained efforts to understand them. For the last fif- teen years or so, I have tried to keep working on these macroscopic phe- nomena. I think I have gained some insights by examining the interrelationships among patterns that have previously been studied in isolation: variation among species in characteristics such as body size, abundance, and range size; spatial variation in local population density within species; and variation in abundance among locally coexisting species. I have also continually pushed myself to come up with mechanistic hypotheses to explain these patterns. This effort has been driven by the premises that for every nonrandom pattern there must be a cause, and that the mechanisms causing such general patterns must be fundamental ecological processes. These kinds of statistical phenomenology are perhaps the most promising clues that nature gives us about where to look for the big, general explanations for the structure and function of complex ecological systems. Many previous investigators, both those who studied these statistical patterns and their colleagues who criticized them, seem to have given up too easily. Many were apparently convinced that the phenomena were simply the manifestation of some kind of randomness. Others seemingly concluded that the explanations were too complex. I have been more optimistic, or at least more stubborn, than most of my colleagues. My optimism may be misplaced, but I hope that others, too, will make sustained efforts to develop and test hypotheses to explain the general statistical patterns.

In my efforts to explain these patterns, I have repeatedly focused more on the form and magnitude of variation than on average trends. I have suggested that many of the statistical distributions can be circumscribed by "constraint envelopes." When the data are plotted on appropriate (usually logarithmic) axes, the constraint envelopes can often be drawn as polygons of distinctive shapes. Drawing each of these constraint lines always forces me to try to come up with a mechanistic hypothesis. I first did this in a paper with my wife on the energetics of territoriality in hummingbirds (Kodric-Brown and Brown 1978). I don't claim that we were the first ones to do this, and others have certainly done so subsequently (e.g., see Lessa and Patton 1989). I do suggest that it is a potentially powerful way to understand how complex systems are influenced by multiple variables.

Another thing that characterizes the macroecological approach is its scale. Macroecology looks at the living world through a macroscope rather than through a microscope, and as a result it sees different things than are revealed by most ecological studies. I must admit to having stumbled serendipitously onto this macroscopic perspective. I was led to it initially by my long-standing interest in biogeography, and later by the practical need to obtain the large number of data points required to characterize statistical patterns. As I began to look through the macroscope, however, I found that it gave me a view of the ecological world that neither my experiments at one study site nor my nonmanipulative comparative studies at a necessarily limited number of field sites could provide. And just as I have obtained insights from my experiments to explain macroecological patterns (e.g., fig. 7.4), so I have been able to apply insights obtained from macroscopic patterns and processes to explain results of the experiments.

Synthesis

It has taken more than two decades to achieve the level of synthesis presented in this book. Initially there was little connection between my microscopic and macroscopic research programs. One was experimental ecology and the other was biogeography. Gradually, however, the connections began to become apparent. I feel that I am beginning to understand how the different levels of biological organization and the different spatial and temporal scales of pattern and process are related to each other.

I have taken two approaches to this synthesis. One is to try to understand how the emergent statistical patterns in populations and communities are influenced by the structure and function of the constituent individual organisms. In particular, I believe that many phenomena at the level of populations, communities, and biotas reflect fundamental constraints on the morphology, physiology, and behavior of the constituent individuals. My background in organismal biology and physiological ecology has been invaluable in formulating mechanistic, individual-based hypotheses. The perspectives of energetics and allometry have been especially useful because of the fundamental importance of thermodynamics and size-dependent scaling processes in all ecological systems.

The second kind of synthesis is an effort to understand how local ecological phenomena affect and are affected by processes operating on larger spatial and temporal scales. This effort has led me to explore the connections between ecology and disciplines such as biogeography, paleontology, and macroevolution. I have tried to break down what I perceive to be artificial boundaries and distinctions. Much of "history" is the result of ecological processes operating in the past. Much of biogeography is the cumulative effect at a regional or continental scale of relationships between organisms and their local environments. But these historical events and biogeographic processes do profoundly influence local ecology. I think that such a synthetic, multiscale perspective is essential if we are to understand in any really deep way the processes that produce and maintain biological diversity on any spatial scale, from local to global, and on any temporal scale, from within a year to the 4.5-billion-year history of life on earth.

Some colleagues (e.g., Wiens 1986, 1989; Ricklefs 1987; Ricklefs and Schluter 1993) have suggested that most traditional ecologists working at small scales have failed to recognize the influence of large-scale processes. I think that this is an oversimplification. The importance of historical events and biogeographic processes has often been recognized, but these phenomena have been viewed as distracting from and complicating efforts to study relationships between organisms and their local environments. Consequently, most ecologists have designed their theories, observations, and experiments so as to hold constant or control for the influence of larger-scale processes. Thus MacArthur (1972a, 257) wrote:

> I predict that there will be erected a two- or three-way classification of organisms and their geometrical and temporal environments, this classification consuming most of the creative energy of ecologists. The future principles of the ecology of coexistence will then be of the form "for organisms of type A, in environments of structure B, such and such relations will hold." This is only a change in emphasis from present ecology. All successful theories, for instance in physics, have initial conditions; with different initial conditions, different things will happen. But I think initial conditions and their classification in ecology will prove to have vastly more effect on outcomes than they do on physics.

There are often good reasons for making such simplifying assumptions and classifications. We must be aware when we do so, however, that we are avoiding some of the most important aspects of ecological complexity. I agree with the above authors, and especially with Roughgarden (e.g., Roughgarden, Gaines, and Pacala 1987; Roughgarden, Gaines, and Possingham 1988), Pulliam (1988), Pimm (1991) and others that we cannot expect to understand the structure and dynamics of local populations and communities without studying the influence of large-scale historical and biogeographic processes. In chapter 2 I pointed out that all ecological systems are inherently (1) open systems, exchanging energy, materials, and organisms with the larger systems in which they are embedded; and (2) historically contingent systems, whose structure and dynamics reflect a continuous process of modification of preexisting systems. These properties of openness and historical contingency are so fundamental to all ecological systems that we must find ways to incorporate them into any synthetic effort to account for the abundance, distribution, and diversity of species.

WHAT IS NEXT?

A New Subdiscipline of Ecology?

It should be obvious that this book only scratches the surface of most subjects. One reviewer of the manuscript remarked that most of the chapters contain enough ideas for at least five doctoral thesis projects or comparable research programs; in a few years it might well be possible to write an entire monograph on the topic of each chapter.

Another reviewer suggested that I end each chapter with a list of suggestions for future research. I have been reluctant to follow this advice, preferring to leave it up to readers to pursue topics that they find interesting. I would be delighted to see every statistical pattern reevaluated with more and better data, with data for different kinds of organisms and environments, and with better statistical and analytical methods. I would like to see every hypothesis and prediction tested, and every one of my tentative conclusions and generalizations challenged. If some of this kind of follow-up and evaluation is done, I expect that it will reveal errors of fact and interpretation. Such errors are not intentional. It is only by making and correcting them that science can advance.

One obvious question is whether macroecology should develop into a recognized, specialized subdiscipline, comparable perhaps to the long-recognized fields of behavioral, physiological, population, community, and ecosystem ecology or the newer one of landscape ecology. Being a holist and a synthesizer, I would be reluctant to advocate such a development. It would detract from what I see as the interdisciplinary nature and the broadly applicable approach of macroecology. I would encourage others to pursue any of the ideas and methods that are of interest, to apply them to any ecological systems, levels of biological organization, and scales of study for which they show promise, and to modify them as needed to answer the questions. I view macroecology mainly as a way of thinking about and investigating complex ecological systems. I hope that others will do likewise.

Identifying Patterns

Rather than suggesting specific topics for investigation, I will mention three areas in which there seem to be especially promising opportunities to extend the macroecological approach. The first is in the identification and characterization of statistical patterns. My macroecological studies have benefited enormously from technological advances, especially from being able to obtain or compile large data bases, from having computers to manage, manipulate, and analyze these data, and from new statistical methods that can be used to perform and evaluate the analyses. My view of the world has been profoundly influenced by the patterns of abundance and distribution revealed by our analyses of the Breeding Bird Survey. The BBS and the statistical tools that we have applied to it have given me a glimpse through a powerful macroscope, allowing me to see the patterns of variation in abundance of different species across the continental landscape and over the last few decades. It is staggering to contemplate how much more research could be done using the BBS and similar data bases.

The analyses that we have done so far are relatively primitive, and our research group is just beginning to exploit methods such as simulation modeling, randomization routines, geostatistics, and Geographic Information Systems (GIS). Further, I suspect that these tools represent only the first stages in the development of techniques for recognizing and characterizing patterns.

Let me illustrate with two examples. Consider the relationship between body mass and population density depicted in figure 5.7. How do we characterize the pattern of variation? I have suggested that it is inadequate to fit an allometric regression equation as Peters (1983) and others have done. Although such a regression is statistically highly significant, it explains only a small proportion of the variation. Brian Maurer and I (e.g., Brown and Maurer 1987, 1989) suggested that much more could be learned by focusing attention on the limits of variation, and we fitted hypothetical constraint lines by eye. Blackburn, Lawton, and Perry (1992) recently developed a statistical method to fit lines to the boundaries of such distributions. But even a rigorously delineated constraint envelope fails to characterize much of the variation, because the data points are not homogeneously distributed within the boundaries. So how can we describe and interpret additional, potentially informative features of the pattern? In addition, since both population density and area of geographic range vary in constrained ways as a function of body size, it would seem informative to plot the joint three-dimensional distribution of these three variables. Brian Maurer and I (unpub.) have done so, but the density of even 300—400 points in three-dimensional space is so low that the pattern was actually less informative than the two-dimensional graphs. However, once the constraint spaces have been delineated and the densities of points within the spaces have been characterized in two dimensions, it should be possible to use the computer to create three-dimensional representations. What form would they take, and would they suggest new ideas?

The second example concerns the spatial variation in abundance within the geographic range of a species. In chapter 4 I describe our recent progress in identifying some of the patterns that are revealed by analysis of BBS data. These analyses, however, only scratch the surface of what might be done using other data bases, GIS, geostatistics, time-space analyses, massively parallel computing, and other approaches. We could use the data that are available from remote sensing and ground-based sources to characterize the abiotic environments of census sites. We could overlay data for different species to search for community-level patterns and possible effects of interspecific interactions on abundance and distribution. Perhaps most interestingly, we could develop a space-time perspective to investigate the joint, interacting patterns of spatial and temporal variation. Because of technological advances, all these possibilities are now or soon will be within our capability.

Testing Hypotheses

Much of my inspiration comes from searching for—and finding—patterns in data. Some of my colleagues are critical of such "fishing expeditions," suggesting that they are bad science. These criticisms might be justified if macroecological research stopped after finding and describing patterns. But every time a pattern is discovered, it cries out for an explanation. Macroecology is as much about the deductive effort of evaluating hypotheses as the inductive effort of searching for patterns.

Here too, there is room for much more work. The evaluation of hypotheses is a long process. It is hard to think of all possible alternative explanations for a pattern, and usually even harder to devise ways to distinguish among the alternatives. Often the alternatives are not mutually exclusive. Only rarely is a single definitive experiment or observation sufficient to give a rigorous test. And it is often impossible to get the explanation of some pattern, which is a manifestation of a complex system, exactly right the first time.[1] Often hypotheses need to be modified rather than simply accepted or rejected.

There are many opportunities both to evaluate more rigorously the hypotheses that I have suggested and to develop and test new ones. Hypotheses are tested by determining whether their assumptions are valid, their arguments are logical, and their predictions are supported by independent empirical observations. Well-designed manipulative experiments offer powerful ways to test both the assumptions and predictions of hypotheses. Unfortunately, experimentation on the scale required to test many macroecological hypotheses is often impractical, unethical, or impossible.[2] There are alternatives. Randomization techniques can often be used to test rigorously the null hypothesis that an apparent pattern would be expected by chance alone, thereby obviating the need to hypothesize some causal mechanism. Computer simulations can often be used to test the logical consequences of proposed mechanisms for the

1. My favorite example is that Darwin, in attempting to explain patterns of diversity such as those he observed in Galápagos finches and tortoises, based his theory of natural selection on the logically faulty process of blending inheritance. Wouldn't it have been counterproductive if Darwin's entire hypothesis had been rejected because one part of it was absolutely wrong?

2. This is by no means always true, however. The recent studies of the ecology and evolution of the "niche" of the bacterium *Escherichia coli* by Lenski and Bennett (1993 and included references) and of the phylogeny of the bacteriophage T7 by Hillis et al. (1992) illustrate how many of the apparent limitations of experiments can be overcome by creatively choosing an appropriate system.

structure and dynamics of complex systems. Often mechanistic hypotheses have additional consequences that can be tested rigorously by making explicit predictions and then evaluating them with independent data. Experimental manipulation is not required to make rigorous tests of predictions. Other disciplines, such as geology and astronomy, have made great advances despite being unable to perform controlled, manipulative experiments.

One note of caution is in order. I think that often ecologists have been too quick to believe that macroecological patterns are just manifestations of random processes. What may seem random at one level might reflect interesting patterns and processes at another, deeper level. I will give two examples. Robert May (1975) pointed out that the approximately lognormal distributions of abundances of coexisting species reported by Preston (1962a,b) and others would be expected if the abundances were determined by the multiplicative interactions of several random variables. Many ecologists misinterpreted May's point to mean that the seemingly general patterns of abundance were simply the result of stochastic processes and hence did not warrant further study.

In a somewhat similar vein, David Wright (1991; see also Hanski and Gyllenberg 1991; Hanski, Kouki, and Halkka 1993) showed that the positive correlation between abundance and distribution among closely related or ecologically similar species (e.g., fig. 4.9) would be expected from a random process. Wright suggested that the Poisson and the negative binomial distributions can be viewed as null models. Tests of the empirical distributions often unequivocally reject the Poisson, but not the negative binomial, which is simply a Poisson modified to include a term for clumping. It would be easy to conclude, as many ecologists seemingly have, that these interspecific relationships between distribution and abundance are relatively trivial consequences of random processes. By extension, the same conclusion could be drawn about intraspecific spatial variation in abundance (see fig. 4.5), which also is often well fit by a negative binomial. But what if we delve deeper, and ask what causes the observed degree of clumping? Much of chapter 4 addresses this question, and a great deal of interesting research remains to be done before it can be answered definitively.

Making Connections to Other Disciplines

In many places in the book, but especially in chapters 10, 11, and 12, I suggest how the approach, ideas, and results of macroecology might be applied to other disciplines such as biogeography, systematics, macroevolution, paleobiology, and conservation biology. Here, too, I have only scratched the surface. My reading and selection of examples has admittedly been spotty; however, I hope that my treatment will be sufficient to suggest some of the possibilities. I will make three points.

First, the emphasis of macroecology on emergent, statistical patterns should be widely applicable. All of these disciplines are trying to study complex, historically contingent, and open systems. All of them are faced with the problem of sorting through an enormous amount of idiosyncratic detail in search of emergent generalities. The statistical approach of macroecology should be applicable in many cases. These disciplines have accumulated large quantities of data: on the species composition of single fossil assemblages, of multiple assemblages at the same place over time, and of multiple assemblages at approximately the same time over geographic space; on the body sizes, trophic relationships, life history characteristics, and other features of fossil organisms; on the reconstructed phylogenetic relationships of many lineages containing many species; on morphological, physiological, behavioral, ecological, and biogeographic characteristics of organisms with well-studied phylogenies; and on the historical geographic ranges of contemporary and fossil organisms. These data, once they have been compiled in standardized data bases, should provide the raw material for searching for seemingly emergent patterns, and then for developing and evaluating hypotheses about the underlying mechanisms.

Second, I think that there are lessons that other disciplines can learn from macroecology. Perhaps one of the most important is not to assume too readily that phenomena are just the consequences of chance events. For example, several macroevolutionary studies suggest that such phenomena as speciation, background and mass extinction, and origin and spread of important evolutionary "innovations" are essentially random. I do not dispute that there may be a substantial probabilistic element in some or all of these phenomena. Nor do I dispute that there may be a significant element of historically contingent uniqueness in them. I do, however, urge my colleagues in other disciplines, like my fellow ecologists, to persevere in the search for pattern and process. For example, Raup, Gould, and colleagues (e.g., Raup et al. 1983; Raup and Gould 1984; Gould 1988) simulated the evolution of lineages in which speciation and extinction events were completely random. Just because these simulations bear some resemblance to real phylogenetic reconstructions, however, does not mean that speciation and extinction are completely stochastic, unpredictable processes (e.g., see Stanley 1973, 1984; Jablonski 1986a; Dial and Marzluff 1989; Martin 1992; Maurer, Brown, and Rusler 1992). Those who study the past should keep in mind Hutchinson's (1965) metaphor: of evolution as a play that takes place on an ecological stage.

Throughout the history of the earth, the environment has played a central role in controlling the abundance, distribution, and diversity of living things and in shaping their characteristics. Some ecologists say that the present holds the keys to understanding the past, and they are right.

Third, ecologists—and microevolutionists—have much to learn from the disciplines that study the history of the earth and its organisms. The historical contingency that characterizes all complex adaptive systems is everywhere apparent in the organization of ecological systems. No understanding of the characteristics of contemporary organisms will be complete without a knowledge of their ancestors. No understanding of contemporary abundance, distribution, and diversity will be complete without a knowledge of the history of the environment and of past ecological relationships. For example, I don't see how any ecologist who studies contemporary terrestrial North American communities could think about these systems in the same way after reading the works of Bernabo and Webb (1977), Martin and Klein (1984), Davis (1986), Delcourt, Delcourt, and Webb (1983), Delcourt and Delcourt (1987), Graham (1986), Betancourt, Van Devender, and Martin (1990), and others on the structure and dynamics of Pleistocene assemblages. Further, because all ecological systems are open systems that exchange energy, materials, and organisms with the larger systems in which they are embedded, no understanding will be complete without a knowledge of relevant phenomena that occur on large spatial scales. Thus I urge my fellow ecologists to make every effort to acquire and apply a historical and biogeographic perspective. Some paleobiologists say that the past holds the keys to understanding the present, and they are equally correct.

CONCLUDING REMARKS

If this book stimulates some of you readers, especially some of the young and highly motivated ones, to pursue some of the questions, try some of the methods, and test some of the hypotheses, it will have served its purpose. I hope that you will find what I have done here interesting and useful. I also hope, however, that you will soon go well beyond where I have left off. I will not be surprised or disappointed when the questions have changed, my methods have been discarded in favor of better ones, and my hypotheses have been proven wrong or inadequate.

More than a decade ago, in my first really macroecological paper (Brown 1981), I said that we should not expect it to be easy to answer Hutchinson's question of why there are so many species. So I am not surprised that this book does not contain the answers. We have made significant progress, but there is still a long way to go. Trying to discover the laws that govern the abundance, distribution, and diversity of species is not for those who are timid or content to do easy science. But for those who are willing to think big, stretch themselves, and persevere, I cannot think of any human endeavor that is more challenging and more fun.

Literature Cited

Allen, T. F. H., and T. W. Hoekstra. 1992. *Toward a unified ecology.* Columbia University Press, New York.

Allen, T. F. H., and T. B. Starr. 1982. *Hierarchy: Perspectives for ecological complexity.* University of Chicago Press. Chicago.

Alvarez, L. W., W. Alvarez, F. Asaro, and H. V. Michel. 1980. Extraterrestrial cause for the Cretaceous-Tertiary extinction. *Science* 208:1095–1108.

Alvarez, W., E. G. Kauffman, F. Surlyk, L. W. Alvarez, F. Asaro, and H. V. Michel. 1984. Impact theory of mass extinctions and the invertebrate fossil record. *Science* 223:1135–41.

Anderson, S. 1977. Geographic ranges of North American terrestrial mammals. *American Museum Novitates* 2629:1–15.

Andrewartha, H. G., and L. C. Birch. 1954. *The distribution and abundance of animals.* University of Chicago Press, Chicago.

Arita, H. T., J. G. Robinson, and K. H. Redford. 1990. Rarity in neotropical forest mammals and its ecological correlates. *Conservation Biology* 4:181–92.

Atkinson, I. A. E., and B. D. Bell. 1973. Offshore and outlying islands. In G. R. Williams, ed., *The natural history of New Zealand,* 372–92. Reed, Wellington.

Basset, Y., and R. L. Kitching. 1991. Species number, species abundance, and body length of arboreal arthropods associated with an Australian rainforest tree. *Ecological Entomology* 16:391–402.

Behrensmeyer, A. K., J. D. Damuth, W. A. DiMichele, R. Potts, H. D. Sues, and S. L. Wing. 1992. *Terrestrial ecosystems through time: Evolutionary paleoecology of terrestrial plants and animals.* University of Chicago Press, Chicago.

Benkman, C. W. 1993. Adaptation to single resources and the evolution of crossbill (*Loxia*) diversity. *Ecological Monographs* 63:305–25.

Bernabo, J. C., and T. Webb III. 1977. Changing patterns in the Holocene pollen record of northeastern North America: A mapped summary. *Quaternary Research* 9:64–96.

Betancourt, J. L., T. R. Van Devender, and P. S. Martin. 1990. *Packrat middens: The last 40,000 years of biotic change.* University of Arizona Press, Tucson.

Blackburn, T. M., J. Lawton, and J. N. Perry. 1992. A method of estimating the slope of upper bounds of plots of body size and abundance for natural assemblages. *Oikos* 65:107–12.

Bock, C. E. 1982. Synchronous fluctuations in Christmas Bird Counts of common Redpolls and Pinon Jays. *Auk* 99:382–83.

———. 1984a. Geographical correlates of abundance vs. rarity in some North American songbirds: A positive correlation. *American Naturalist* 122:295–99.

———. 1984b. Geographical correlates of rarity vs. abundance in some North American winter landbirds. *Auk* 101:266–73.

———. 1987. Distribution-abundance relationships of some North American landbirds: A matter of scale? *Ecology* 68:124–29.

Bock, C. E., and L. W. Lepthien. 1972. Winter eruptions of Red-breasted Nuthatches in North America, 1950–1970. *American Birds* 26:558–61.

———. 1976. Synchronous eruption of boreal seed-eating birds. *American Naturalist* 110:559–71.

Bock, C. E., and R. E. Ricklefs. 1983. Range size and local abundance of some North American songbirds: A positive correlation. *American Naturalist* 122:295–99.

Bohning-Gaese, K., M. L. Taper, and J. H. Brown. 1993. Are declines in North American insectivorous songbirds due to causes on the breeding range? *Conservation Biology* 7:76–86.

Boltzmann, L. 1905. *Populare schriften.* J. A. Barth, Leipzig.

Bonner, J. T. 1988. *The evolution of complexity by means of natural selection.* Princeton University Press, Princeton, N.J.

Boucot, A. J. 1976. Rates of size increase and phyletic evolution. *Nature* 261:694–95.

Bowers, M. A., and J. H. Brown. 1982. Body size and coexistence in desert rodents: Chance or community structure. *Ecology* 63:391–400.

Boyd, E. M., and S. A. Nunneley. 1964. Banding records substantiating the changed status of 10 species of birds since 1900 in the Connecticut Valley. *Bird-Banding* 35:1–8.

Brooks, D. L. 1985. Historical ecology: A new approach to studying the evolution of ecological associations. *Annals of the Missouri Botanical Garden* 72:660–80.

Brooks, D. L., and D. A. McLennan. 1991. *Phylogeny, ecology, and behavior.* University of Chicago Press, Chicago.

Brown, J. H. 1968. *Adaptation to environmental temperature in two species of woodrats,* Neotoma cinerea *and* N. albigula. Miscellaneous Publications, Museum of Zoology, University of Michigan 135:1–48.

———. 1971a. Mammals on mountaintops: Nonequilibrium insular biogeography. *American Naturalist* 105:467–78.

———. 1971b. Mechanisms of competitive exclusion between two species of chipmunks *Eutamias. Ecology* 52:306–11.

———. 1975. Geographical ecology of desert rodents. In M. L. Cody and J. M. Diamond, eds., *Ecology and evolution of communities,* 315–41. Harvard University Press, Cambridge, Mass.

———. 1981. Two decades of homage to Santa Rosalia: Toward a general theory of diversity. *American Zoologist* 21:877–88.

————. 1984. On the relationship between abundance and distribution of species. *American Naturalist* 124:255–79.

————. 1986. Two decades of interaction between the MacArthur-Wilson model and the complexities of mammalian distributions. *Biological Journal of the Linnean Society* 28:231–51.

————. 1988. Species diversity. In A. Myers and R. S. Giller, eds., *Analytical biogeography*, 57–89. Chapman and Hall, London.

————. 1989. Patterns, modes and extents of invasions by vertebrates. In J. A. Drake, H. A. Mooney, F. di Castri, R. H. Groves, F. J. Kruger, M. Rejmanek, and M. Williamson, eds., *Biological invasions*, 85–109. John Wiley & Sons, New York.

————. 1994. Complex ecological systems. In G. A. Cowan, D. Pines, and D. Melzer, eds., *Complexity: Metaphors, models, and reality*, 419–49. Santa Fe Institute Studies in the Science of Complexity, Proceedings Volume XVIII. Addison-Wesley, Reading, Mass.

————. In press. Organisms and species as complex adaptive systems: Linking the biology of populations with the physics of ecosystems. In C. G. Jones and J. H. Lawton, eds., *Linking species and ecosystems: Essays on the conjunction of perspectives*. Chapman and Hall, New York.

Brown, J. H., and D. W. Davidson. 1977. Competition between seed-eating rodents and ants in desert ecosystems. *Science* 196:880–82.

Brown, J. H., D. W. Davidson, J. C. Munger, and R. S. Inouye. 1986. Experimental community ecology: The desert granivore system. In J. Diamond and T. J. Case, eds., *Community ecology*, 41–61. Harper & Row, New York.

Brown, J. H., and C. R. Feldmeth. 1971. Evolution in constant and fluctuating environments: Thermal tolerances of desert pupfish *Cyprinodon*. *Evolution* 25:390–98.

Brown, J. H., and A. C. Gibson. 1983. *Biogeography*. Mosby, St. Louis, Mo.

Brown, J. H., and E. J. Heske. 1990a. Control of a desert-grassland transition by a keystone rodent guild. *Science* 250:1705–7.

————. 1990b. Temporal changes in a Chihuahuan Desert rodent community. *Oikos* 59:290–302.

Brown, J. H., and A. Kodric-Brown. 1977. Turnover rates in insular biogeography: Effect of immigration on extinction. *Ecology* 58:445–49.

Brown, J. H., and M. A. Kurzius. 1987. Composition of desert rodent faunas: Combinations of coexisting species. *Annales Zoologici Fennici* 24:227–37.

————. 1989. Spatial and temporal variation in guilds of North American desert rodents. In D. W. Morris, Z. Abramsky, B. J. Fox, and M. R. Willig, eds., *Patterns in the structure of mammalian communities*, 71–90. Special Publication no. 28, The Museum, Texas Tech University, Lubbock.

Brown, J. H., and M. V. Lomolino. 1989. Independent discovery of the equilibrium theory of island biogeography. *Ecology* 70:1954–57.

Brown, J. H., P. A. Marquet, and M. L. Taper. 1993. Evolution of body size: Consequences of an energetic definition of fitness. *American Naturalist* 142:573–84.

Brown, J. H., and B. A. Maurer. 1986. Body size, ecological dominance, and Cope's rule. *Nature* 324:248–50.

————. 1987. Evolution of species assemblages: Effects of energetic constraints and species dynamics on the diversification of North American avifauna. *American Naturalist* 130:1–17.

————. 1989. Macroecology: The division of food and space among species on continents. *Science* 243:1145–50.

Brown, J. H., and J. C. Munger. 1985. Experimental manipulation of a desert rodent community: Food addition and species removal. *Ecology* 66:1545–63.

Brown, J. H., and P. F. Nicoletto. 1991. Spatial scaling of species composition: Body masses of North American land mammals. *American Naturalist* 138:1478–1512.

Brown, J. H., and Z. Zeng. 1989. Comparative population ecology of eleven species of rodents in the Chihuahuan Desert. *Ecology* 70:1507–25.

Brown, W. L. 1957. Centrifugal speciation. *Quarterly Review of Biology* 32:247–77.

Brown, W. L, and E. O. Wilson. 1956. Character displacement. *Systematic Zoology* 5:49–64.

Burgman, M. A. 1989. The habitat volumes of scarce and ubiquitous plants: A test of the model of environmental control. *American Naturalist* 133:228–39.

Bush, G. L. 1969. Sympatric host race formation and speciation in frugivorous flies of the genus *Rhagoletis* (*Diptera, Tephritidae*). *Evolution* 23:237–51.

————. 1975. Modes of animal speciation. *Annual Review of Ecology and Systematics* 6:339–64.

Bystrak, D. 1979. The breeding bird survey. *Sialia* 1:74–79, 87.

Calder, W. A. III. 1984. *Size, function, and life history.* Harvard University Press, Cambridge, Mass.

————. 1989. Avian longevity and aging. In D. E. Harrison, ed., *Genetic effects on aging,* vol. 2, 185–204. Telford Press, Caldwell, N.J.

Carlquist, S. 1965. *Island life.* Natural History Press, Garden City, N.Y.

Carson, H. L. 1968. The population flush and the genetic consequences. In R. C. Lewontin, ed., *Population biology and evolution,* 123–27. Syracuse University Press, Syracuse, N.Y.

Carson, H. L., and K. Y. Kaneshiro. 1976. Drosophila of Hawaii: Systematics and ecological genetics. *Annual Review of Ecology and Systematics* 7:311–45.

Charlesworth, B., R. Lande, and M. Slatkin. 1982. A neo-Darwinian commentary on macroevolution. *Evolution* 36:474–98.

Charnov, E. L. 1992. On evolution of the age of maturity. *Evolutionary Ecology* 6:307–11.

Clements, F. E. 1916. *Plant succession: An analysis of the development of vegetation.* Publication no. 242. Carnegie Institute, Washington, D.C.

————. 1949. *Dynamics of vegetation.* Hafner Press, New York.

Cody, M. L., and H. A. Mooney. 1978. Convergence versus nonconvergence in Mediterranean-climate ecosystems. *Annual Review of Ecology and Systematics* 9:265–321.

Collins, S. L., and S. M. Glenn. 1990. A hierarchical analysis of species' abundance patterns in grassland vegetation. *American Naturalist* 135:633–48.

Connell, J. H. 1961a. The effects of competition, predation by *Thais lapillus,* and other factors on natural populations of the barnacle, *Balanus balanoides*. *Ecological Monographs* 31:61–104.

———. 1961b. The influence of interspecific competition and other factors on the distribution of the barnacle *Chthamalus stellatus*. *Ecology* 42:410–23.

———. 1975. Some mechanisms producing structure in natural communities: A model and evidence from field experiments. In M. L. Cody and J. M. Diamond, eds., *Ecology and evolution of communities*, 460–90. Harvard University Press, Cambridge, Mass.

Connor, E. F., S. H. Faeth, D. Simberloff, and P. A. Opler. 1980. Taxonomic isolation and the accumulation of herbivorous insects: A comparison of introduced and native trees. *Ecological Entomology* 5:205–11.

Connor, E. F., and D. Simberloff. 1979. The assembly of species communities: Chance or competition? *Ecology* 60:1132–40.

Cook, R. E. 1969. Variation in species density of North American birds. *Systematic Zoology* 18:63–84.

Cotgreave, P., and P. H. Harvey. 1992. Relationships between body size, abundance, and phylogeny in bird communities. *Functional Ecology* 6:248–56.

Cowan, G. A., D. Pines, and D. Melzer, eds. 1994. *Complexity: Metaphors, models, and reality*. Santa Fe Institute Studies in the Science of Complexity, Proceedings Volume XVIII. Addison-Wesley, Reading, Mass.

Cox, G. W., and R. E. Ricklefs. 1977. Species diversity, ecological release, and community structuring in Caribbean land bird faunas. *Oikos* 29:60–66.

Cracraft, J. 1989. Speciation and its ontology: The empirical consequences of alternative species concepts for understanding patterns and processes of differentiation. In D. Otte and J. A. Endler, eds., *Speciation and its consequences*, 27–59. Sinauer Associates, Sunderland, Mass.

———. 1990. The origin of evolutionary novelties: Pattern and process at different hierarchial levels. In M. H. Nitecki, ed., *Evolutionary innovations*, 21–44. University of Chicago Press, Chicago.

Croizat, L. 1958. *Panbiogeography*. Published by the author, Caracas.

Damuth, J. 1981. Population density and body size in mammals. *Nature* 290:699–700.

———. 1987. Interspecific allometry of population density in mammals and other animals: The independence of body mass and population energy-use. *Biological Journal of the Linnean Society* 31:193–246.

———. 1991. Of size and abundance. *Nature* 351:268–69.

Danielson, B. J. 1991. Communities in a landscape: The influence of habitat heterogeneity on interactions between species. *American Naturalist* 138:1105–20.

Darwin, C. 1859. *On the origin of species*. John Murray, London.

Davis, M. B. 1986. Climatic instability, time lags, and community disequilibrium. In J. Diamond and T. J. Case, eds., *Community ecology*, 269–84. Harper & Row, New York.

Dawson, W. R., and C. Carey. 1976. Seasonal acclimatization to temperature in cardueline finches. *Journal of Comparative Physiology* 112:317–33.

Dayan T., and D. Simberloff. 1994. Morphological relationships among coexisting heteromyids: An incisive dental character. *American Naturalist* 143:462–77.

Dayan, T., D. Simberloff, E. Tchernov, and Y. Yom-Tov. 1989. Inter- and intraspecific character displacement in mustelids. *Ecology* 70:1526–39.

———. 1990. Feline canines: Community-wide character displacement in the small cats of Israel. *American Naturalist* 136:39–60.

———. 1992. Canine carnassials: Character displacement among the wolves, jackals, and foxes of Israel. *Biological Journal of the Linnean Society* 45:315–31.

Dayan, T., E. Tchernov, Y. Yom-Tov, and D. Simberloff. 1989. Ecological character displacement in Saharo-Arabian *Vulpes:* Outfoxing Bergmann's rule. *Oikos* 55:263–72.

DeAngelis, D. L. In press. Relationships between the energetics of species and the thermodynamics of ecosystems. In C. G. Jones and J. H. Lawton, eds., *Linking species and ecosystems: Essays on the conjunction of perspectives.* Chapman and Hall, New York.

Delcourt, H. R., P. A. Delcourt, and T. Webb III. 1983. Dynamic plant ecology: The spectrum of vegetational change in space and time. *Quarterly Scientific Review* 1:153–75.

Delcourt, P. A., and H. R. Delcourt. 1987. *Long-term forest dynamics of the temperate zone.* Springer-Verlag, Berlin.

Dial, K. P., and J. M. Marzluff. 1988. Are the smallest organisms the most diverse? *Ecology* 69:1620–24.

———. 1989. Nonrandom diversification within taxonomic assemblages. *Systematic Zoology* 38:26–37.

Diamond, J. M. 1975. Assembly of species communities. In M. L. Cody and J. M. Diamond, eds., *Ecology and evolution of communities,* 342–444. Harvard University Press, Cambridge, Mass.

———. 1984. Historic extinctions: A Rosetta stone for understanding prehistoric extinctions. In P. S. Martin and R. G. Klein, eds., *Quaternary extinctions,* 824–62. University of Arizona Press, Tucson.

Diamond, J. M., and C. R. Veitch. 1981. Extinctions and introductions in the New Zealand avifauna: Cause and effect? *Science* 211:499–501.

Dice, L. R. 1947. *Effectiveness of selection by owls of deer mice* (Peromyscus maniculatus) *which contrast in color with their background.* Contributions from the Laboratory of Vertebrate Biology, University of Michigan, no. 34.

Dice, L. R., and P. M. Blossom 1937. *Studies of mammalian ecology in southwestern North America with special attention to the colors of desert mammals.* Publication no. 485. Carnegie Institute, Washington, D.C.

Dobson, F. S., and J. Yu. In press. Rarity in neotropical forest mammals revisited. *Conservation Biology.*

Dobzhansky, T. 1950. Evolution in the tropics. *American Scientist* 38:209–11.

Dodd, A. P. 1959. The biological control of prickly pear in Australia. In A. Keast, ed., *Biogeography and ecology in Australia.* Monographiae Biologicae 8. Junk, The Hague.

Donoghue, M. J. 1989. Phylogenies and the analysis of evolutionary sequences, with examples from seed plants. *Evolution* 43:1137–56.

Doyle, J. A. 1978. Origin of angiosperms. *Annual Review of Ecology and Systematics* 9:365–92.

Doyle, J. A., and M. J. Donoghue. 1986. Seed plant phylogeny and the origin of angiosperms: An experimental cladistic approach. *Botanical Review* 52:321–431.

———. 1987. The importance of fossils in elucidating seed plant phylogeny and macroevolution. *Review of Paleobotany and Palynology* 50:63–95.

Drake, J. A., H. A. Mooney, F. di Castri, R. H. Groves, F. J. Kruger, M. Rejmanek, and M. Williamson, eds. 1989. *Biological invasions.* John Wiley & Sons, New York.

Dueser, R. D., J. H. Porter, and J. L. Dooley Jr. 1989. Direct tests for competition in North American rodent communities: Synthesis and prognosis. In D. W. Morris, Z. Abramsky, B. J. Fox, and W. R. Willig, eds., *Patterns in the structure of mammalian communities,* 105–25. Texas Tech University Press, Lubbock.

Dunham, A., B. W. Grant, and K. L. Overall. 1989. Interfaces between biophysical and physiological ecology and the population ecology of terrestrial vertebrate ectotherms. *Physiological Zoology* 62(2):335–55.

Dunning, J. B. Jr., and J. H. Brown. 1982. Summer rainfall and winter sparrow densities: A test of the food limitation hypothesis. *Auk* 99:123–29.

Ehleringer, J. R. 1984a. Ecology and ecophysiology of leaf pubescence in North American desert plants. In E. Rodriguez, P. Healey, and I. Mehta, eds., *Biology and chemistry of plant trichomes,* 113–32. Plenum Press, New York.

Ehleringer, J. R. 1984b. Interspecific competitive effects on water relations, growth, and reproduction in *Encelia farinosa. Oecologia* 63:153–58.

Ehrlich, P. R., and P. H. Raven. 1965. Butterflies and plants: A study in coevolution. *Evolution* 18:586–608.

Eisenberg, J. F. 1981. *The mammalian radiations.* University of Chicago Press, Chicago.

Eldredge, N. 1979. Alternative approaches to evolutionary theory. *Bulletin of the Carnegie Museum of Natural History* 13:7–19.

———. 1985. *Unfinished synthesis: Biological hierarchies and modern evolutionary thought.* Oxford University Press, Oxford.

———. 1989. *Macroevolutionary dynamics.* McGraw-Hill, New York.

Eldredge, N., and S. J. Gould. 1972. Punctuated equilibria: An alternative to phyletic gradualism. In T. J. M. Schopf, ed., *Models in paleobiology,* 82–115. Freeman, Cooper & Co., San Francisco.

Erickson, R. O. 1945. The *Clematis fremontii* var. *riehlii* population in the Ozarks. *Annals of the Missouri Botanical Garden* 32:413–60.

Erwin, T. L. 1983a. Beetles and other insects of tropical rain forest canopies at Manaus, Brazil, sampled by insecticidal fogging. In S. L. Sutton, T. C. Whitmore, and A. C. Chadwick, eds., *Tropical rainforest ecology and management,* 59–75. Blackwell, Edinburgh.

———. 1983b. Tropical forest canopies: The last biotic frontier. *Bulletin of the Entomological Society of America* 29:14–19.

Farlow, J. O. 1993. On the rareness of big, fierce animals: Speculations about the body sizes, population densities, and geographic ranges of predatory mammals and large carnivorous dinosaurs. *American Journal of Science* 293-A:167–99.

Feder, J. 1988. *Fractals.* Plenum Press, New York and London.

Feldmeth, C. R. 1981. The evolution of thermal tolerance in desert pupfish (genus *Cyprinodon*). In R. J. Naiman and D. L. Soltz, eds., *Fishes in North American deserts,* 357–84. John Wiley & Sons, New York.

Feldmeth, C. R., E. A. Stone, and J. H. Brown. 1974. Increased scope for thermal tolerance upon acclimating pupfish (*Cyprinodon*) to cycling temperatures. *Journal of Comparative Physiology* 89:39–44.

Felsenstein, J. 1985. Phylogenies and the comparative method. *American Naturalist* 125:1–15.

Findley, J. S. 1993. *Bats: A community perspective.* Cambridge University Press, Cambridge.

Fischer, A. G. 1960. Latitudinal variation in organic diversity. *Evolution* 14:64–81.

Fisher, R. A., A. S. Corbet, and C. B. Williams. 1943. The relationship between the number of species and the number of individuals in a random sample of an animal population. *Journal of Animal Ecology* 12:42–58.

Forman, R. T. T., and M. Godron. 1986. *Landscape ecology.* Wiley, New York.

Fox, B. J. 1989. Small-mammal community pattern in Australian heathlands: A taxonomically based rule for species assembly. In D. W. Morris, Z. Abramsky, B. J. Fox, and M. R. Willig, eds., *Patterns in the structure of mammalian communities,* 91–103. Special Publication no. 28. The Museum, Texas Tech University, Lubbock.

Fox, B. J., and J. H. Brown. 1993. Assembly rules for functional groups in North American desert rodent communities. *Oikos* 67:358–70.

Fox, B. J., and G. L. Kirkland. 1992. An assembly rule for functional groups applied to North American soricid communities. *Journal of Mammalogy* 73:491–503.

Fretwell, S. D. 1972. *Populations in a seasonal environment.* Princeton University Press, Princeton, N.J.

Fretwell, S. D., and H. L. Lucas. 1970. On territorial behavior and other factors influencing habitat distribution in birds. *Acta Biotheoretica* 19:16–36.

Frey, J. K. 1992. Response of a mammalian faunal element to climatic changes. *Journal of Mammalogy* 73:43–50.

Fryer, G., and T. D. Iles. 1972. *The cichlid fishes of the Great Lakes of Africa: Their biology and evolution.* Oliver & Boyd, Edinburgh.

Fuentes, E. R. 1976. Ecological convergence of lizard communities in Chile and California. *Ecology* 57:3–17.

Futuyma, D. J., and S. S. McCafferty. 1990. Phylogeny and the evolution of host plant associations in the leaf beetle genus *Ophreaella* (Coleoptera, Chrysomelidae). *Evolution* 44:1885–1913.

Futuyma, D. J., and M. Slatkin, eds. 1983. *Coevolution.* Sinauer Associates, Sunderland, Mass.

Gaston, K. J., and J. H. Lawton. 1988a. Patterns in body size, population dynamics, and regional distributions of bracken herbivores. *American Naturalist* 132:622–80.

———. 1988b. Patterns in the distribution and abundance of insect populations. *Nature* 331:709–12.

———. 1989. Insect herbivores on bracken do not support the core-satellite hypothesis. *American Naturalist* 134:761–77.

————. 1990a. Effects of scale and habitat on the relationship between regional distribution and local abundance. *Oikos* 58:329–35.

————. 1990b. The population ecology of rare species. *Journal of Fish Biology* 37:97–104.

Gause, G. F. 1934. *The struggle for existence.* Williams & Wilkins, Baltimore.

Gilbert, B. M., and L. D. Martin. 1984. Late Pleistocene fossils of Natural Trap Cave, Wyoming, and the climatic model of extinction. In P. S. Martin and R. G. Klein, eds., *Quaternary extinctions,* 138–47. University of Arizona Press, Tucson.

Gilbert, F. S. 1980. The equilibrium theory of island biogeography: Fact or fiction? *Journal of Biogeography* 7:209–35.

Gilpin, M., and I. Hanski. 1991. *Metapopulation dynamics.* Academic Press, London.

Gittelman, J. L. 1985. Carnivore body size: Ecological and taxonomic correlates. *Oecologia* 67:540–54.

Gittelman, J. L., and H. K. Luh. 1992. On comparing comparative methods. *Annual Review of Ecology and Systematics* 23:383–404.

Glazier, D. S. 1980. Ecological shifts and the evolution of geographically restricted species of North American *Peromyscus* (mice). *Journal of Biogeography* 7:63–83.

————. 1988. Temporal variability of abundance and the distribution of species. *Oikos* 47:309–14.

Gleason, H. A. 1917. The structure and development of the plant association. *Bulletin of the Torrey Botanical Club* 44:463–81.

————. 1926. The individualistic concept of plant association. *Bulletin of the Torrey Botanical Club* 53:7–26.

Goel, N. S., and N. Richter-Dyn. 1974. *Stochastic models in biology.* Academic Press, New York.

Goodman, D. 1986. The demography of chance extinction. In M. E. Soulé, ed., *Viable populations,* 11–34. Cambridge University Press, Cambridge.

Gotelli, N. J., and D. Simberloff. 1987. The distribution and abundance of tallgrass prairie plants: A test of the core-satellite hypothesis. *American Naturalist* 130:18–35.

Gould, S. J. 1980. Is a new and general theory of evolution emerging? *Paleobiology* 6:119–30.

————. 1988. Presidential Address—Trends as changes in variance: A new slant on progress and directionality in evolution. *Journal of Paleontology* 62:319–29.

Gould, S. J., and N. Eldredge. 1977. Punctuated equilibrium: The tempo and mode of evolution reconsidered. *Paleobiology* 3:115–51.

Gould, S. J., D. M. Raup, J. J. Sepkoski Jr., T. J. M. Schopf, and D. S. Simberloff. 1977. The shape of evolution: A comparison of real and random clades. *Paleobiology* 3:23–40.

Graham, R. W. 1986. Responses of mammalian communities to environmental changes during the late Quaternary. In J. Diamond and T. J. Case, eds., *Community ecology,* 300–313. Harper & Row, New York.

Graham, R. W., and E. L. Lundelius. 1984. Coevolutionary disequilibrium. In

P. S. Martin and R. G. Klein, eds., *Quaternary extinctions*, 223–49. University of Arizona Press, Tucson.

Grajal, A., S. D. Strahl, R. Parra, M. G. Dominguez, and A. Neher. 1989. Foregut fermentation in the hoatzin, a neotropical leaf-eating bird. *Science* 245:1236–38.

Grant, P. R. 1986. *Ecology and evolution of Darwin's Finches*. Princeton University Press, Princeton, N.J.

Grayson, D. K. 1977. Pleistocene avifaunas and the overkill hypothesis. *Science* 195:691–93.

———. 1984. Explaining Pleistocene extinctions. In P. S. Martin and R. G. Klein, eds., *Quaternary extinctions*, 807–23. University of Arizona Press, Tucson.

———. 1987. The biogeographic history of small mammals in the Great Basin: Observations on the last 20,000 years. *Journal of Mammalogy* 68:359–75.

Grayson, D. K., and S. D. Livingston. 1993. Missing mammals on Great Basin mountains: Holocene extinctions and inadequate knowledge. *Conservation Biology* 7:527–32.

Grinnell, J. 1914. An account of the mammals and birds of the lower Colorado River Valley. *University of California Publications in Zoology* 12:51–294.

———. The niche-relationships of the California Thrasher. *Auk* 34:427–33.

———. 1922. The role of the "accidental." *Auk* 39:373–80.

Grinnell, J., and R. T. Orr. 1934. Systematic review of the *Californicus* group of the rodent genus *Peromyscus*. *Journal of Mammalogy* 15:210–17.

Grinnell, J., and H. S. Swarth. 1913. An account of the birds and mammals of the San Jacinto area of southern California. *University of California Publications in Zoology* 10:197–406.

Guilday, J. E. 1984. Pleistocene extinction and environmental change. In P. S. Martin and R. G. Klein, eds., *Quaternary extinctions*, 250–58. University of Arizona Press, Tucson.

Hansen, T. A. 1980. Influence of larval dispersal and geographic distribution on species longevity in neo-gastropods. *Paleobiology* 6:193–207.

Hanski, I. 1982a. Communities of bumblebees: Testing the core-satellite hypothesis. *Annales Zoologici Fennici* 19:65–73.

———. 1982b. Distributional ecology of anthropochorus plants in villages surrounded by forest. *Annales Botanici Fennici* 19:1–15.

———. 1982c. Dynamics of regional distribution: The core and satellite species hypothesis. *Oikos* 38:210–21.

———. 1991. Single-species metapopulation dynamics: Concepts, models and observations. In M. Gilpin and I. Hanski, eds., *Metapopulation dynamics*, 17–38. Academic Press, London.

Hanski, I., and M. Gyllenberg. 1991. *Two general metapopulation models and the core-satellite species hypothesis*. Lulea University of Technology, Department of Applied Mathematics, Research Report 4. Lulea, Sweden.

Hanksi, I., J. Kouki, and A Halkka. 1993. Three explanations of the positive relationship between distribution and abundance of species. In R. E. Ricklefs and D. Schluter, eds., *Species diversity in ecological communities*, 108–16. University of Chicago Press, Chicago.

Harestad, A. S., and F. L. Bunnell. 1979. Home range and body weight: A reevaluation. *Ecology* 60:389–402.

Harris, P., D. Peschken, and J. Milroy. 1969. The status of biological control of the weed *Hypericum perforatum* in British Columbia. *Canadian Entomology* 101:1–15.

Harvey, P. H., and H. C. J. Godfray. 1987. How species divide resources. *American Naturalist* 129:318–20.

Harvey, P. H., and S. Nee. 1991. How to live like a mammal. *Nature* 350:23–24.

Harvey, P. H., and M. D. Pagel. 1991. *The comparative method in evolutionary biology*. Oxford University Press, Oxford.

Hastings, A., C. L. Hom, S. Ellner, P. Turchin, and H. C. J. Godfray. 1993. Chaos in ecology: Is mother nature a strange attractor? *Annual Review of Ecology and Systematics* 24:1–33.

Hayward, J. S. 1965a. Metabolic rate and its temperature-adaptive significance in six geographic races of *Peromyscus*. *Canadian Journal of Zoology* 43:309–23.

———. 1965b. Microclimate temperature and its adaptive significance in six geographic races of *Peromyscus*. *Canadian Journal of Zoology* 43:341–50.

Heaney, L. R. 1984a. Climatic influences on life-history tactics and behavior of North American tree squirrels. In J. O. Murie and G. R. Michener, eds., *The biology of ground-dwelling squirrels*, 43–78. University of Nebraska Press, Lincoln.

———. 1984b. Mammalian species richness on islands on the Sunda Shelf, Southeast Asia. *Oecologia* 61:11–17.

Hengeveld, R. 1989. *Dynamics of biological invasions*. Chapman and Hall, London.

———. 1990. *Dynamic biography*. Cambridge University Press, Cambridge.

Hengeveld, R., and J. Haeck. 1981. The distribution of abundance. II. Models and implications. *Proceedings of the Koninklijke Nederlandse Akademie van Wetenschappen*, series C, *Biological and Medical Sciences* 84:257–84.

———. 1982. The distribution of abundance. I. Measurements. *Journal of Biogeography* 9:303–16.

Hennig, W. 1966. *Phylogenetic systematics*. University of Illinois Press, Urbana.

Hershkovitz, P. 1966. Mice, land bridges, and Latin America faunal interchange. In R. L. Wenzel and V. J. Tipton, eds., *Parasites of Panama*, 725–47. Field Museum of Natural History, Chicago.

Heske, E. J., J. H. Brown, and S. Mistry. 1994. Long-term experimental study of a Chihuahuan desert rodent community: 13 years of competition. *Ecology* 75:438–45.

Hill, N. P., and J. M. Hagan III. 1991. Population trends of some northeastern North American landbirds: A half-century of data. *Wilson Bulletin* 103:165–82.

Hillis, D. M., J. J. Bull, M. E. White, M. R. Badgett, and I. J. Molineux. 1992. Experimental phylogenetics: Generation of a known phylogeny. *Science* 255:589–92.

Holling, C. S. 1992. Cross-scale morphology, geometry, and dynamics of ecosystems. *Ecological Monographs* 62:447–502.

Holt, R. D. 1983. Immigration and the dynamics of peripheral populations. In A.

Roden and K. Miyata, eds., *Advances in herpetology and evolutionary biology*, 680–94. Museum of Comparative Zoology, Harvard University, Cambridge, Mass.

———. 1987. Population dynamics and evolutionary processes: The manifold roles of habitat selection. *Evolutionary Ecology* 1:331–47.

———. 1992. A neglected facet of island biogeography: The role of internal spatial dynamics in area effects. *Theoretical Population Biology* 41:354–71.

Holt, R. D., and M. S. Gaines. 1992. The analysis of adaptation in heterogeneous landscapes: Implications for the evolution of fundamental niches. *Evolutionary Ecology* 6:433–47.

Hopf, F. A., and J. H. Brown. 1986. The bullseye method for testing for randomness in ecological communities. *Ecology* 67:1139–55.

Hopf, F. A., T. J. Valone, and J. H. Brown. 1993. Competition theory and the structure of ecological communities. *Evolutionary Ecology* 7:142–54.

Horn, M. H., and L. G. Allen. 1978. A distributional analysis of California coastal marine fishes. *Journal of Biogeography* 5:23–42.

Horner, E. B. 1954. *Arboreal adaptations of* Peromyscus, *with special reference to use of the tail.* Contributions from the Laboratory of Vertebrate Biology, University of Michigan, no. 61.

Huelsenbeck, J. P., and D. M. Hillis. 1993. Success of phylogenetic methods in the four-taxon case. *Systematic Biology* 42:247–64.

Humphries, C. J., and L. R. Parenti. 1986. *Cladistic biogeography.* Clarendon Press, Oxford.

Hutchinson, G. E. 1957. Concluding remarks. *Cold Spring Harbor Symposia on Quantitative Biology* 22:415–27.

———. 1959. Homage to Santa Rosalia, or why are there so many kinds of animals? *American Naturalist* 93:145–59.

———. 1965. *The ecological theater and the evolutionary play.* Yale University Press, New Haven, Conn.

Hutchinson, G. E., and R. H. MacArthur. 1959. A theoretical ecological model of size distributions among species of animals. *American Naturalist* 93:117–25.

Hutto, R. L. 1989. The effect of habitat alteration on migratory land birds in a West Mexican tropical deciduous forest: A conservation perspective. *Conservation Biology* 3:138–48.

Jablonski, D. 1985. Marine regressions and mass extinctions: A test using the modern biota. In J. W. Valentine, ed., *Phanerozoic diversity patterns: Profiles in macroevolution*, 335–54. Princeton University, Princeton, N.J.

———. 1986a. Causes and consequences of mass extinctions: A comparative approach. In D. K. Elliot, ed., *Dynamics of extinction*, 183–229. Wiley, New York.

———. 1986b. Larval ecology and macroevolution in marine invertebrates. *Bulletin of Marine Science* 39(2):565–87.

———. 1987. Heritability at the species level: Analysis of geographic ranges of cretaceous mollusks. *Science* 238:360–63.

Jablonski, D., and R. A. Lutz. 1983. Larval ecology of marine benthic invertebrates: Paleobiological implications. *Biological Reviews* 58:21–89.

James, F. C., R. F. Johnston, N. O. Wamer, G. J. Niemi, and W. J. Boecklen.

1984. The Grinnellian niche of the Wood Thrush. *American Naturalist* 124:17–47.

Janzen, D. H. 1967. Why mountain passes are higher in the tropics. *American Naturalist* 101:233–49.

——. 1982. Differential seed survival and passage rates in cows and horses, surrogate Pleistocene dispersal agents. *Oikos* 38:150–56.

——. 1985. On ecological fitting. *Oikos* 45:308–10.

Johnson, A. R., J. A. Wiens, B. T. Milne, and T. O. Crist. 1992. Animal movements and population dynamics in heterogeneous landscapes. *Landscape Ecology* 7(1):63–75.

Karasov, W. H. 1990. Digestion in birds: Chemical and physiological determinants and ecological implications. *Studies in Avian Biology* 13:391–415.

Kareiva, P. M., and M. Andersen. 1988. Spatial aspects of species interactions: The wedding of models and experiments. In A. Hastings, ed., *Community ecology*, 38–54. Springer-Verlag, New York.

Kareiva, P. M., J. G. Kingsolver, and R. B. Huey, eds.. 1992. *Biotic interactions and global change*. Sinauer Associates, Sunderland, Mass.

Kauffman, S. A. 1993. *The origins of order.* Oxford University Press, New York.

Kiester, A. R. 1971. Species density of North American amphibians and reptiles. *Systematic Zoology* 20:127–37.

Kodric-Brown, A., and J. H. Brown. 1978. Influence of economics, interspecific competition, and sexual dimorphism on territoriality of migrant Rufous Hummingbirds. *Ecology* 59:285–96.

——. 1993a. Highly structured fish communities in Australian desert springs. *Ecology* 74:1847–55.

——. 1993b. Incomplete data sets in community ecology and biogeography: A cautionary tale. *Ecological Applications* 3:736–42.

La Barbera, M. 1986. The evolution and ecology of body size. In D. M. Raup and D. Jablonski, eds., *Patterns and processes in the history of life*, 69–98. Springer-Verlag, Berlin.

La Brecque, M. 1992. To model the otherwise unmodelable. *Mosaic* 23:12–23.

Lack, D. 1947. *Darwin's Finches.* Cambridge University Press, Cambridge.

Laurance, W. F., and E. Yensen. 1985. Rainfall and winter sparrow densities: A view from the northern Great Basin. *Auk* 102:152–58.

Lawton, J. H. 1990. Species richness and population dynamics of animal assemblages. Patterns in body size: Abundance and space. *Philosophical Transactions of the Royal Society of London*, series B, 330:283–91.

Lazell, J. D. Jr. 1966. Studies on *Anolis reconditus* Underwood and Williams. *Bulletin of the Institute of Jamaica* 18, part I.

Leigh, E. 1975. Population fluctuations and community structure. In W. H. Van Dobben and R. H. Lowe-McConnell, eds., *Unifying concepts in ecology*, 67–88. Junk, The Hague.

Lenski, R. E., and A. F. Bennett. 1993. Evolutionary response of *Escherichia coli* to thermal stress. *American Naturalist* 142:S47-S64.

Lessa, E. P., and J. L. Patton. 1989. Structural constraints, recurrent shapes, and allometry in pocket gophers (genus *Thomomys*). *Biological Journal of the Linnean Society* 36:349–63.

Levin, S. A. 1992. The problem of pattern and scale in ecology. *Ecology* 73:1943–67.

Levins, R. 1969. Some demographic and genetic consequences of environmental heterogeneity for biological control. *Bulletin of the Entomological Society of America* 15:237–40.

Lewin, R. 1992. *Complexity.* Macmillan, New York.

Liem, K. F. 1973. Evolutionary strategies and morphological innovations: Cichlid pharyngeal jaws. *Systematic Zoology* 22:425–41.

Lindemann, R. L. 1942. The trophic-dynamic aspect of ecology. *Ecology* 23:399–418.

Lister, A. M. 1989. Rapid dwarfing of red deer on Jersey in the last interglacial. *Nature* 342:539–42.

Lockwood, J. L., and M. P. Moulton. 1994. Ecomorphological pattern in Bermuda birds: The influence of competition and implications for nature preserves. *Evolutionary Ecology* 8:53–60.

Lockwood, J. L., M. P. Moulton, and S. K. Anderson. 1993. Morphological assortment and the assembly of communities of introduced Passeriformes on oceanic islands: Tahiti versus Oahu. *American Naturalist* 141:389–408.

Lomolino, M. V. 1985. Body sizes of mammals on islands: The island rule reexamined. *American Naturalist* 125:310–16.

———. 1986. Mammalian community structure on islands: Immigration, extinction, and interactive effects. *Biological Journal of the Linnean Society* 28:1–21.

———. 1989. Bioenergetics of cross-ice movements by *Microtus pennsylvanicus, Peromyscus leucopus,* and *Blarina brevicauda. Holarctic Ecology* 12:213–18.

———. 1994a. Immigrations and distribution patterns of insular mammals: Studying fundamental processes in island biogeography. *Ecography* 16:376–79.

———. 1994b. Species richness of mammals inhabiting nearshore archipelagoes: Area, isolation and immigration filters. *Journal of Mammalogy* 75:39–49.

Lomolino, M. V., J. C. Creighton, G. D. Schnell, and D. L. Certain. In press. Ecology and conservation of the endangered American burying beetle (*Nicrophorus americanus*). *Conservation Biology.*

Lotka, A. J. 1922. Contribution to the energetics of evolution. *Proceedings of the National Academy of Science* 8:147–55.

———. 1925. *Principles of physical biology.* Williams & Wilkins, Baltimore.

Lubchenco, J., and B. Menge. 1978. Community development and persistence in a low rocky intertidal zone. *Ecological Monographs* 59:67–94.

Lubchenco, J., A. O. Olson, L. B. Brubaker, S. R. Carpenter, M. M. Holland, S. P. Hubbell, S. A. Levin, J. A. MacMahon, P. A. Matson, J. M. Melillo, H. A. Mooney, C. H. Peterson, H. R. Pulliam, L. A. Real, P. J. Regal, and P. G. Risser. 1991. The sustainable biosphere initiative: An ecological research agenda. *Ecology* 72:371–412.

MacArthur, R. H. 1957. On the relative abundance of bird species. *Proceedings of the National Academy of Sciences U.S.A.* 43:293–95.

———. 1958. Population ecology of some warblers of northeastern coniferous forests. *Ecology* 39:599–619.

———. 1960a. On the relation between reproductive value and optimal predation. *Proceedings of the National Academy of Sciences U.S.A.* 46:144–45.

———. 1960b. On the relative abundance of species. *American Naturalist* 94:25–36.

———. 1961. Population effects of natural selection. *American Naturalist* 95:195–99.

———. 1962. Some generalized theorems of natural selection. *Proceedings of the National Academy of Sciences U.S.A.* 48:1893–97.

———. 1965. Ecological consequences of natural selection. In T. H. Waterman and H. J. Morowitz, eds., *Theoretical and mathematical biology*, 388–97. Blaisdell, New York.

———. 1966. Note on Mrs. Pielou's comments. *Ecology* 47:1074.

———. 1972a. Coexistence of species. In J. Behnke, ed., *Challenging biological problems,* 253–59. Oxford University Press, New York.

———. 1972b. *Geographical ecology: Patterns in the distribution of species.* Harper & Row, New York.

MacArthur, R. H., and E. O. Wilson. 1963. An equilibrium theory of insular zoogeography. *Evolution* 17:373–87.

———. 1967. *The theory of island biogeography.* Princeton University Press, Princeton, N.J.

Malthus, R. T. 1798. *An essay on the principle of population as it affects the future improvement of society.* Johnson, London.

Mandelbrot, B. B. 1983. *The fractal geometry of nature.* Freeman, San Francisco.

Marquet, P. A., S. A. Navarrete, and J. C. Castilla. 1990. Scaling population density to body size in rocky intertidal communities. *Science* 250:1061–1184.

Marshall, L. G. 1979. A model for paleobiogeography of South American cricetine rodents. *Paleobiology* 5:126–32.

Marshall, L. G., S. D. Webb, J. J. Sepkoski, and D. M. Raup. 1982. Mammalian evolution and the great American interchange. *Science* 215:1351–57.

Martin, P. S. 1966. Africa and Pleistocene overkill. *Nature* 212:339–42.

———. 1967. Prehistoric overkill. In P. S. Martin and H. E. Wright Jr., eds., *Pleistocene extinctions: The search for a cause,* 75–120. Yale University Press, New Haven, Conn.

———. 1984a. Catastrophic extinctions and late Pleistocene blitzkrieg: Two radiocarbon dates. In M. H. Nitecki, ed., *Extinctions,* 153–89. University of Chicago Press, Chicago.

———. 1984b. Prehistoric overkill: The global model. In P. S. Martin and R. G. Klein, eds., *Quaternary extinctions,* 354–403. University of Arizona Press, Tucson.

———. 1990. Who or what destroyed our mammoths? In L. D. Agenbroad, J. I. Meal, and L. W. Nelson, eds., *Megafauna and man: Discovery of America's heartland,* 109–18. Hot Springs, South Dakota.

Martin, P. S., and R. G. Klein, eds. 1984. *Quaternary extinctions.* University of Arizona Press, Tucson.

Martin, R. A. 1979. Fossil history of the rodent genus *Sigmodon. Evolutionary Monographs* 2:1–36.

————. 1984. The evolution of cotton rat body mass. In H. H. Genoways and M. R. Dawson, eds., *Contributions in Quaternary paleontology: A volume in memorial to John E. Guilday*, 252–66. Special Publication no. 8, Carnegie Museum of Natural History, Pittsburgh, Pa.

————. 1986. Energy, ecology and cotton rat evolution. *Paleobiology* 12:370–82.

————. 1992. Generic species richness and body mass in North American mammals: Support for the inverse relationship of body size and speciation rate. *Historical Biology* 6:73–90.

Marzluff, J. M., and K. P. Dial. 1991. Life history correlates of taxonomic diversity. *Ecology* 72:428–39.

Maurer, B. A. 1994. *Geographic population analysis*. Blackwell Scientific, Oxford.

————. In press. Analysis of range fragmentation in geographic populations of North American birds. In W. K. Michener, J. W. Brunt, and S. G. Stafford, eds., *Environmental information management and analysis: Ecosystem to global scales*. Taylor and Francis, London.

Maurer, B. A., and J. H. Brown. 1988. Distribution of biomass and energy use among species of North American terrestrial birds. *Ecology* 69:1923–32.

Maurer, B. A., J. H. Brown, and R. D. Rusler. 1992. The micro and macro in body size evolution. *Evolution* 46(4):939–53.

Maurer, B. A., and M. Villard. In press. Geographic variation in abundance of North American birds. *Research and Exploration*.

May, R. M. 1975. Patterns of species abundance and diversity. In M. L. Cody and J. M. Diamond, eds., *Ecology and evolution of communities*, 81–120. Harvard University Press, Cambridge, Mass.

————. 1976. The evolution of ecological systems. *Scientific American* 239(3):160–75.

————. 1978. The dynamics and diversity of insect faunas. In L. A. Mound and N. Waloff, eds., *Diversity of insect faunas*, 188–204. Blackwell, Oxford.

————. 1986. The search for patterns in the balance of nature: Advances and retreats. *Ecology* 67:1115–26.

————. 1988. How many species are there on earth? *Science* 241:1441–49.

Mayr, E. 1942. *Systematics and the origin of species*. Columbia University Press, New York.

————. 1963. *Animal species and evolution*. Harvard University Press, Cambridge, Mass.

McDonald, K. A., and J. H. Brown. 1992. Using montane mammals to model extinctions due to global change. *Conservation Biology* 6:409–15.

McDonnell, M. J., and S. T. A. Pickett. 1990. Ecosystem structure and function along urban-rural gradients: An unexploited opportunity for ecology. *Ecology* 71:1232–37.

McIntosh, R. P. 1985. *The background of ecology*. Cambridge University Press, Cambridge.

McMahon, T. 1973. Size and shape in biology. *Science* 179:1201–4.

McMahon, T., and J. T. Bonner. 1983. *On size and life*. Scientific American Books, New York.

Menge, B. A, E. L. Berlow, C. A. Blanchette, S. A. Navarette, and S. B. Yamada.

In press. The keystone species concept: Variation in interaction strength in a rocky intertidal habitat. *Ecological Monographs*.

Miller, R. R. 1966. Geographical distribution of Central American freshwater fishes. *Copeia* 4:773–802.

Milne, B. T. 1988. Measuring the fractal geometry of landscapes. *Applied Mathematics and Computation* 27:67–79.

———. 1991. Heterogeneity as a multiscale characteristic of landscapes. In J. Kolasa and S. T. A. Pickett, eds., *Ecological heterogeneity*. Springer-Verlag, Berlin.

———. 1992. Spatial aggregation and neutral models in fractal landscapes. *American Naturalist* 139:32–57.

Milne, B. T., M. G. Turner, J. A. Wiens, and A. R. Johnson. 1992. Interactions between the fractal geometry of landscapes and allometric herbivory. *Theoretical Population Biology* 41:337–53.

Molles, M. C., and C. N. Dahm. 1990. El Niño, La Niña, and North American stream ecology. *Journal of the North American Benthological Society* 9: 68–76.

Mooney, H. A. 1977. *Convergent evolution in Chile and California*. Dowden, Hutchinson, and Ross, Stroudsburg, Pa.

Mooney, H. A., E. R. Fuentes, and B. I. Kronberg. 1993. *Earth system responses to global change*. Academic Press, San Diego, Calif.

Mooney, H. A., J. Kummerow, A. W. Johnson, D. J. Parsons, S. Keeley, A. Hoffman, R. I. Hays, J. Giliberto, and C. Chu. 1977. The producers—their resources and adaptive responses. In H. A. Mooney, ed., *Convergent evolution in Chile and California*, 85–143. Dowden, Hutchinson, and Ross, Stroudsburg, Pa.

Morris, D. W. In press. Habitat matching: Alternatives and implications to population and communities. *Evolutionary Ecology*.

Morse, D. R., N. E. Stock, and J. H. Lawton. 1988. Species numbers, species abundance, and body length relationships of arboreal beetles in Bornean lowland rain forest trees. *Ecological Entomology* 13:25–37.

Morton, E. S. 1978. Avian arboreal folivores: Why not? In G. G. Montgomery, ed., *The ecology of arboreal folivores*, 123–30. Smithsonian Institution Press, Washington, D.C.

Motomura, I. 1932. A statistical treatment of associations (in Japanese). *Japanese Journal of Zoology* 44:379–83.

Moulton, M. P., and S. L. Pimm. 1986. The extent of competition in shaping an introduced avifauna. In J. Diamond and T. J. Case, eds., *Community ecology*, 80–97. Harper & Row, New York.

———. 1987. Morphological assortment in introduced Hawaiian passerines. *Evolutionary Ecology* 1:113–24.

Mulligan, H. W., ed. 1970. *The African trypanosomiases*. Wiley-Interscience, New York.

Munroe, E. G. 1948. The geographical distribution of butterflies in the West Indies. Ph.D. dissertation, Cornell University, Ithaca, N.Y.

———. 1953. The size of island faunas. In *Proceedings of the Seventh Pacific*

Science Congress of the Pacific Science Association, vol. 4, *Zoology,* 52–53. Whitcome and Tombs, Auckland, New Zealand.

Myers, A. A., and P. S. Giller, eds. 1989. *Analytical biogeography.* Chapman and Hall, London.

Myers, N. 1987. The extinction spasm impending: Synergisms at work. *Conservation Biology* 1:14–21.

Nagy, K. A. 1987. Field metabolic rate and food requirement scaling in mammals and birds. *Ecological Monographs* 57:111–28.

Nee, S., A. F. Read, J. J. D. Greenwood, and P. H. Harvey. 1991. The relationship between abundance and body size in British birds. *Nature* 351:312–13.

Nelson, G., and N. Platnick. 1981. *Systematics and biogeography: Cladistics and vicariance.* Columbia University Press, New York.

Nitecki, M. H., ed. 1984. *Extinctions.* University of Chicago Press, Chicago.

Nobel, P. S. 1980. Morphology, surface temperatures, and northern limits of columnar cacti in the Sonoran Desert. *Ecology* 61:1–7.

Odum, E. P. 1959. *Fundamentals of ecology.* 2d ed. Saunders, Philadelphia.

———. 1969. The strategy of ecosystem development. *Science* 164:262–70.

Odum, H. T. 1983. *Systems ecology: An introduction.* John Wiley & Sons, New York.

Odum, H. T., and R. C. Pinkerton. 1955. Times speed regulator: The optimum efficiency for maximum power output in physical and biological systems. *American Scientist* 43:331–43.

Olson, S. L. 1989. Extinction on islands: Man as a catastrophe. In D. Western and M. Pearl, eds., *Conservation for the twenty-first century,* 50–53. Oxford University Press, New York.

Olson, S. L., and H. F. James. 1982a. Fossil birds from the Hawaiian Islands: Evidence for wholesale extinction by man before Western contact. *Science* 217:633–35.

———. 1982b. *Promodromus of the fossil avifauna of the Hawaiian islands.* Smithsonian Contributions to Zoology no. 365. Smithsonian Institution Press, Washington, D.C.

———. 1984. The role of Polynesians in the extinction of the avifauna of the Hawaiian Islands. In P. S. Martin and R. G. Klein, eds., *Quaternary extinctions,* 768–80. University of Arizona Press, Tucson.

O'Neill, R. V., D. L. DeAngelis, J. B. Waide, and T. F. H. Allen. 1987. *A hierarchical concept of ecosystems.* Princeton University Press, Princeton, N.J.

Orians, G. H. 1980. Micro and macro in ecological theory. *BioScience* 30:79.

Otte, D. 1981. *The North American grasshoppers.* Harvard University Press, Cambridge, Mass.

Owen-Smith, R. N. 1987. Pleistocene extinctions: The pivotal role of megaherbivores. *Paleobiology* 13:351–62.

———. 1988. *Megaherbivores: The influence of very large body size on ecology.* Cambridge University Press, Cambridge.

———. 1989. Megafaunal extinctions: The conservation message from 11,000 B. C. *Conservation Biology* 3:405–11.

Pagel, M. D., and P. H. Harvey. 1988. Recent developments in the analysis of comparative data. *Quarterly Review of Biology* 63:413–40.

————. 1989. Comparative methods for examining adaptation depend on evolutionary models. *Folia Primatologica* 53:203–20.

Pagel, M. D., P. H. Harvey, and H. C. J. Godfray. 1991. Species-abundance, biomass, and resource-use distributions. *American Naturalist* 138:836–50.

Pagel, M. D., R. M. May, and A. R. Collie. 1991. Ecological aspects of the geographical distribution and diversity of mammalian species. *American Naturalist* 137:791–815.

Paine, R. T. 1966. Food web complexity and species diversity. *American Naturalist* 100:65–75.

————. 1974. Intertidal community structure: Experimental studies on the relationship between a dominant competitor and its principal predator. *Oecologia* 15:93–120.

Patterson, B. D. 1987. The principle of nested subsets and its implications for biological conservation. *Conservation Biology* 1:323–34.

————. 1990. On the temporal development of nested subset patterns of species composition. *Oikos* 59:330–42.

Patterson, B. D., and W. Atmar. 1986. Nested subsets and the structure of insular mammalian faunas and archipelagos. *Biological Journal of the Linnean Society* 28:65–82.

Peters, R. H. 1976. Tautology in evolution and ecology. *American Naturalist* 110:1–12.

————. 1983. *The ecological implications of body size.* Cambridge University Press, Cambridge.

Peters, R. H. 1991. *A critique for ecology.* Cambridge University Press, Cambridge.

Peters, R. H., and T. E. Lovejoy, eds. 1992. *Global warming and biological diversity.* Yale University Press, New Haven, Conn.

Peters, R. H., and J. V. Raelson. 1984. Relations between individual size and mammalian population density. *American Naturalist* 124:498–517.

Pielou, E. C. 1966. Comment on a report by J. H. Vandermeer and R. H. MacArthur concerning the broken stick model of species abundance. *Ecology* 47:1073–74.

————. 1977. Mathematical ecology. 2d ed. Harper & Row, New York.

Pimm, S. L. 1991. *The balance of nature? Ecological issues in the conservation of species and communities.* University of Chicago Press, Chicago.

Pimm, S. L., H. L. Jones, and J. Diamond. 1988. On the risk of extinction. *American Naturalist* 132:757–85.

Pimm, S. L., and A. Redfearn. 1988. The variability of population densities. *Nature* 334:613–14.

Porter, W. P., and D. M. Gates. 1969. Thermodynamic equilibria of animals with the environment. *Ecological Monographs* 39:245–70.

Pough, F. H. 1980. The advantages of ectothermy for tetrapods. *American Naturalist* 115:92–112.

————. 1983. Amphibians and reptiles as low-energy systems. In W. P. Aspey and S. I. Lustick, eds., *Behavioral energetics,* 141–88. Ohio State University Press, Columbus.

Preston, F. W. 1948. The commonness, and rarity, of species. *Ecology* 29:254–83.

————. 1962a. The canonical distribution of commonness and rarity: Part I. *Ecology* 43:185–215.

————. 1962b. The canonical distribution of commonness and rarity: Part II. *Ecology* 43:410–32.

Pulliam, H. R. 1983. Ecological community theory and the coexistence of sparrows. *Ecology* 64:45–52.

————. 1988. Sources, sinks, and population regulation. *American Naturalist* 132:652–61.

Pulliam, H. R., and D. J. Danielson. 1991. Sources, sinks, and habitat selection: A landscape perspective on population dynamics. *American Naturalist* 137:550–60.

Pulliam, H. R., and G. S. Mills. 1977. The use of space by wintering sparrows. *Ecology* 58:1393–99.

Pulliam, H. R., and T. H. Parker. 1979. Population regulation of sparrows. *Fortschritte de Zoologie* 25:137–47.

de Queiroz, D., and J. Gauthier. 1992. Phylogenetic taxonomy. *Annual Review of Ecology and Systematics* 23:449–80.

Quinn, J. F. 1983. Mass extinctions in the fossil record. *Science* 219:1239–40.

Rabinowitz, D. 1981. Seven forms of rarity. In J. Synge, ed., *The biological aspects of rare plant conservation*, 205–17. Wiley, Chichester.

Rabinowitz, D., S. Cairns, and T. Dillon. 1986. Seven forms of rarity and their frequency in the flora of the British Isles. In M. E. Soulé, ed., *Conservation biology: The science of scarcity and diversity*, 182–204. Sinauer Associates, Sunderland, Mass.

Rapoport, E. H. 1982. *Aerography: Geographical strategies of species.* Pergamon, Oxford.

Rapoport, E. H., M. E. Diaz Betancourt, and I. R. Lopez Moreno. 1983. *Aspectos de la ecologia urbana en la Cuidad de Mexico: Flora de las calles y baldios.* Editorial Limusa, Mexico, D.F.

Raup, D. M. 1979. Size of the Permo-Triassic bottleneck and its evolutionary implications. *Science* 206:217–18.

————. 1986. Biological extinction in earth history. *Science* 231:1528–33.

Raup, D. M, and S. J. Gould. 1984. Stochastic simulation and evolution of morphology: Towards a nomothetic paleontology. *Systematic Zoology* 23:305–22.

Raup, D. M., S. J. Gould, T. J. M. Schopf, and D. Simberloff. 1983. Stochastic models of phylogeny and the evolution of diversity. *Journal of Geology* 81:525–42.

Raup, D. M., and D. Jablonski. 1986. *Patterns and processes in the history of life.* Springer-Verlag, Berlin.

Raup, D. M., and J. J. Sepkoski Jr. 1982. Mass extinctions in the marine fossil record. *Science* 215:1501–3.

————. 1984. Periodicities of extinctions in the geological past. *Proceedings of the National Academy of Sciences U.S.A.* 81:801–5.

Reig, O. A. 1978. Roedores cricetidos del plioceno superior de la Provincia Buenos Aires (Argentina). *Publication de la Museum de Cience National, Mara del Plata* 2:164–90.

————. 1989. Karyotypic repatterning as one triggering factor in cases of explosive speciation. In A. Fondevila, ed., *Evolutionary biology of transient unstable populations,* 246–89. Springer-Verlag, Berlin.

Reiss, J. O. 1989. *The allometry of growth and reproduction.* Cambridge University Press, Cambridge.

Repasky, R. R. 1991. Temperature and the northern distributions of wintering birds. *Ecology* 72:2274–85.

Rice, W. R., and G. W. Salt. 1990. The evolution of reproductive isolation as a correlated character under sympatric conditions: Experimental evidence. *Evolution* 44:1140–52.

Ricklefs, R. E. 1987. Community diversity: Relative roles of local and regional processes. *Science* 235:167–71.

Ricklefs, R. E., and G. W. Cox. 1972. Taxon cycles in the West Indian avifauna. *American Naturalist* 106:195–219.

Ricklefs, R. E., and D. Schluter, eds. 1993. *Species diversity in ecological communities.* University of Chicago Press, Chicago.

Robbins, C. S., D. Bystrak, and P. H. Geissler. 1986. *The breeding bird survey: Its first fifteen years, 1965–1979.* U.S. Fish and Wildlife Service, Resource Publication 157. Washington, D.C.

Robbins, C. S., J. R. Sauer, R. S. Greenberg, and S. Droege. 1989. Population declines in North American birds that migrate to the Neotropics. *Proceedings of the National Academy of Sciences U.S.A.* 86:7658–62.

Rohde, K., M. Heap, and D. Heap. 1993. Rapoport's rule does not apply to marine teleosts and cannot explain latitudinal gradients in species richness. *American Naturalist* 142:1–16.

Root, R. B. 1967. The niche exploitation pattern of the Blue-gray Gnatcatcher. *Ecological Monographs* 37:317–50.

Root, T. 1988a. *Atlas of wintering North American birds—an analysis of Christmas Bird Count data.* University of Chicago Press, Chicago.

———. 1988b. Energy constraints on avian distributions and abundances. *Ecology* 69:330–39.

———. 1988c. Environmental factors associated with avian distributional boundaries. *Journal of Biogeography* 15:489–505.

Rosenzweig, M. L. 1975. On continental steady states of species diversity. In M. L. Cody and J. M. Diamond, eds., *Ecology and evolution of communities,* 121–40. Harvard University Press, Cambridge, Mass.

———. 1978. Competitive speciation. *Biological Journal of the Linnean Society* 10:275–89.

———. 1992. Species diversity gradients: We know more and less than we thought. *Journal of Mammalogy* 73:715–30.

Roth, V. L. 1990. Insular dwarf elephants: A case study in body mass estimation and ecological inference. In J. Damuth and B. J. MacFadden, eds., *Body size in mammalian paleobiology: Estimation and biological implications,* 151–79. Cambridge University Press, New York.

Roughgarden, J. 1979. *Theory of population genetics and evolutionary ecology: An introduction.* Macmillan, New York.

Roughgarden, J., S. D. Gaines, S. W. Pacala. 1987. Supply side ecology: The role of physical transport processes. In J. H. R. Gee and P. S. Giller, eds., *Organization of communities: Past and present*, 491–518. Blackwell, Oxford.

Roughgarden, J., S. D. Gaines, and H. Possingham. 1988. Recruitment dynamics in complex life cycles. *Science* 241:1460–66.

Roughgarden, J., D. Heckel, and E. R. Fuentes. 1983. Coevolutionary theory and the biogeography and community structure of *Anolis*. In R. B. Huey, E. R. Pianka, and T. W. Schoener, eds., *Lizard ecology: Studies of a model organism*, 371–410. Harvard University Press, Cambridge, Mass.

Salt, G. W., ed. 1983. A roundtable on research in ecology and evolutionary biology. *American Naturalist* 122:583–705.

Salthe, S. N. 1985. *Evolving hierarchical systems: Their structure and representation*. Columbia University Press, New York.

Schaffer, W. M., and M. Kot. 1986a. Chaos in ecological systems: The coals that Newcastle forgot. *Trends in Ecology and Evolution* 1:58–63.

———. 1986b. Differential systems in ecology and epidemiology. In A. V. Holden, ed., *Chaos*, 158–78. Princeton University Press, Princeton, N.J.

Schluter, D. 1984. Morphological and phylogenetic relations among Darwin's finches. *Evolution* 38:921–30.

Schmidt-Nielsen, K. 1984. *Scaling, why is animal size so important?* Cambridge University Press, Cambridge.

Schneider, E. D., and J. J. Kay. In press. Life as a manifestation of the second law of thermodynamics. *The Journal of Advances in Math and Computers in Medicine*.

Schoener, T. W. 1968. Sizes of feeding territories among birds. *Ecology* 49:123–31.

———. 1970. Size patterns in West Indian *Anolis* lizards. II. Correlations with the sizes of particular sympatric species—displacement and convergence. *American Naturalist* 104:155–74.

———. 1984. Size differences among sympatric, bird-eating hawks: A worldwide survey. In D. R. Strong Jr., D. Simberloff, L. G. Abele, and A. B. Thistle, eds., *Ecological communities: Conceptual issues and the evidence*, 254–81. Princeton University Press, Princeton, N.J.

———. 1987. The geographical distribution of rarity. *Oecologia* 74:161–73.

———. 1988. The ecological niche. In J. M. Cherrett, ed., *Ecological concepts*, 79–113. Blackwell Scientific, Oxford.

———. 1990. The geographical distribution of rarity: Misinterpretation of atlas methods affects some empirical conclusions. *Oecologia* 82:567–68.

Schoener, T. W., and D. A. Spiller. 1987. High population persistence in a system with high turnover. *Nature* 330:474–77.

Schoener, T. W., and D. A. Spiller. 1992. Is extinction rate related to temporal variability in population size? An empirical answer for orb spiders. *American Naturalist* 139:1176–1207.

Schrodinger, E. 1946. *What is life?* Cambridge University Press, Cambridge.

Scott, J. M., F. Davis, B. Csuti, R. Noss, B. Butterfield, C. Groves, H. Anderson, S. Caicco, F. D'Erchia, T. C. Edwards Jr., J. Ulliman, and R. G. Wright. 1993. Gap analysis; a geographic approach to protection of biological diversity. Wildlife Monographs no. 123.

Sepkoski, J. J. Jr. 1978. A kinetic model of Phanerozoic taxonomic diversity. I. Analysis of marine orders. *Paleobiology* 4:223–51.

———. 1982. Mass extinctions in the Phanerozoic oceans: A review. Geological Society of America, Special Paper 190. Geological Society of America, Boulder, Colo.

———. 1984. A kinetic model of Phanerozoic taxonomic diversity. III. Post-Paleozoic families and mass extinctions. *Paleobiology* 10:246–67.

———. 1986. Global bioevents and the question of periodicity. In O. Walliser, ed., *Lecture Notes in Earth Sciences*, vol 8, *Global Bio-Events*. Springer-Verlag, Berlin.

Simberloff, D. 1974. Equilibrium theory of island biogeography and ecology. *Annual Review of Ecology and Systematics* 5:161–82.

———. 1980. Dynamic equilibrium island biogeography: The second stage. In R. Nohring, ed., *Acta XVII Congressus Internationalis Ornithologici*, vol. 2, 1289–95. Verlag der Deutschen Ornithologen-Gessellschaft, Berlin.

———. 1986. Are we on the verge of a mass extinction in tropical rain forests? In D. K. Elliot, ed., *Dynamics of extinction*, 165–80. John Wiley & Sons, New York.

Simberloff, D., and W. Boecklen. 1981. Santa Rosalia reconsidered: Size ratios and competition. *Evolution* 35:1206–28.

———. 1991. Patterns of extinction in the introduced Hawaiian avifauna: A re-examination of the role of competition. *American Naturalist* 138:300–327.

Simberloff, D., and T. Dayan. 1991. The guild concept and the structure of ecological communities. *Annual Review of Ecology and Systematics* 22:115–43.

Simpson, G. G. 1964. Species density of North American Recent mammals. *Evolution* 15:413–46.

———. 1980. *Splendid isolation: The curious history of mammals in South America*. Yale University Press, New Haven, Conn.

Smith, C. C. 1970. The coevolution of pine squirrels (*Tamiasciurus*) and conifers. *Ecological Monographs* 40:349–71.

Smith, F. E. 1961. Density dependence in the Australian thrips. *Ecology* 42:403–7.

———. 1963. Density-dependence. *Ecology* 44:220.

Sober, E. 1984. *The nature of selection*. University of Chicago Press, Chicago.

Soltz, D. L., and R. J. Naiman. 1978. *The natural history of native fishes in the Death Valley system*. Natural History Museum of Los Angeles County, Los Angeles, Calif.

Soulé, M. E. 1980. Threshold for survival: Maintaining fitness and evolutionary potential. In M. E. Soulé and B. A. Wilcox, eds., *Conservation biology: An evolutionary-ecological perspective*, 151–70. Sinauer Associates, Sunderland, Mass.

———, ed. 1986. *Conservation biology: The science of scarcity and diversity*. Sinauer Associates, Sunderland, Mass.

Soulé, M. E., and B. A. Wilcox, eds. 1980. *Conservation biology: An evolutionary-ecological perspective*. Sinauer Associates, Sunderland, Mass.

Southwood, T. R. E. 1978. The components of diversity. In L. A. Mound and N. Waloff, eds., *The diversity of insect faunas*. Blackwell, Oxford.

Southwood, T. R. E., V. C. Moran, and C. E. D. Kennedy. 1982. The richness, abundance, and biomass of the arthropod communities on trees. *Journal of Animal Ecology* 51:635–49.

Stanley. S. M. 1973. An explanation for Cope's rule. *Evolution* 27:1–26.

———. 1979. *Macroevolution: Pattern and process.* Freeman, San Francisco.

———. 1982. Macroevolution and the fossil record. *Evolution* 36:460–73.

———. 1984. Marine mass extinctions: A dominant role for temperature. In M. H. Nitecki, ed., *Extinctions,* 69–117. University of Chicago Press, Chicago.

———. 1986. Population size, extinction, and speciation: The fission effect in Neogene Bivalvia. *Paleobiology* 12:89–110.

Steadman, D. W. 1989. Extinction of birds in eastern Polynesia: A review of the records, and comparisons with other Pacific island groups. *Journal of Archaeological Science* 16:177–205.

Steadman, D. W., and P. S. Martin. 1984. Extinction of birds in the late Pleistocene. In P. S. Martin and R. G. Klein, eds., *Quaternary extinctions,* 466–77. University of Arizona Press, Tucson.

Stebbins, G. L., and F. J. Ayala. 1981. Is a new evolutionary synthesis necessary? *Science* 213:967–71.

Steenbergh, W. F., and C. H. Lowe. 1976. Ecology of the saguaro. I. The role of freezing weather in a warm-desert plant population. In *Research in the parks,* 49–92. National Park Service Symposium, serial no. 1. U.S. Government Printing Office, Washington, D.C.

———. 1977. Ecology of the saguaro. II. Reproduction, germination, establishment, growth, and survival of the young plant. National Park Service Science Monograph, serial no. 8. U.S. Government Printing Office, Washington, D.C.

Stehli, F. G. 1968. Taxonomic diversity gradients in pole locations: The recent model. In E. T. Drake, ed., *Evolution and environment,* 163–227. Peabody Museum Centennial Symposium. Yale University Press, New Haven, Conn.

Stehli, F. G., R. G. Douglas, and N. D. Newell. 1969. Generation and maintenance of gradients in taxonomic diversity. *Science* 164:947–49.

Stehli, F. G., and J. W. Wells. 1971. Diversity and age patterns in hermatypic corals. *Systematic Zoology* 20:115–26.

Stevens, G. C. 1989. The latitudinal gradient in geographic range: How so many species coexist in the tropics. *American Naturalist* 133:240–56.

———. 1992. The elevational gradient in altitudinal range: An extension of Rapoport's latitudinal rule to altitude. *American Naturalist* 140:893–911.

Stevens, G. C., and J. F. Fox. 1991. The causes of treeline. *Annual Review of Ecology and Systematics* 22:177–91.

Strong, D. R. Jr. 1974. Rapid asymptotic species accumulation in phytophagous insect communities: The pests of Cacao. *Science* 185:1064–66.

Strong, D. R. Jr., E. D. McCoy, and J. R. Rey. 1977. Time and the number of herbivore species: The pests of sugar cane. *Ecology* 58:167–75.

Strong, D. R., D. Simberloff, L. G. Abele, and A. B. Thistel, eds. 1984. *Ecological communities: Conceptual issues and the evidence.* Princeton University Press, Princeton, N.J.

Stuart, A. J. 1991. Mammalian extinctions in the late Pleistocene of northern Eurasia and North America. *Biological Reviews* 66:453–562.

Sugihara, G. 1980. Minimal community structure: An explanation of species abundance patterns. *American Naturalist* 116:770–87.

———. 1989. How do species divide resources? *American Naturalist* 133:458–63.

Sumner, F. B. 1932. Genetic, distributional, and evolutionary studies of the subspecies of deer mice *Peromyscus maniculatus*. *Bibliography of Genetics* 9:1–106.

Swisher, C. C. III, J. M Grajales-Nishimura, A. Montanari, S. V. Margolis, P. Claeys, W. Alvarez, P. Renne, E. Cedillo-Pardo, F. J-M. R. Maurrasse, G. H. Curtis, J. Smit, and M. O. McWilliams. 1992. Coeval 40AR/39AR ages of 65.0 million years ago from Chicxulub crater melt rock and Cretaceous-Tertiary boundary tektites. *Science* 257:954–58.

Taylor, C. M., and N. J. Gotelli. In press. The macroecology of *Cyprinella:* Correlates of phylogeny, body size, and geographic range. *American Naturalist.*

Templeton, A. R. 1980. Theory of speciation via the founder principle. *Genetics* 94:1011–38.

Terborgh, J. 1989. *Where have all the birds gone?* Princeton University Press, Princeton, N.J.

Thompson, D. B. 1990. Different spatial scales of adaptation in the climbing behavior of *Peromyscus maniculatus:* Geographic variation, natural selection, and gene flow. *Evolution* 44:952–65.

Thompson, D. W. 1917. *On growth and form.* Cambridge University Press, Cambridge.

Tilman, D. 1989. Ecological experimentation: Strengths and conceptual problems. In G. E. Likens, ed., *Long-term studies in ecology,* 136–57. Springer-Verlag, New York.

Tonn, W. M., J. J. Magnuson, M. Rask, and J. Toivonen. 1990. Intercontinental comparison of small-lake fish assemblages: Balance between local and regional processes. *American Naturalist* 136:345–75.

Townsend, C. R., and P. Calow. 1981. *Physiological ecology: An evolutionary approach to resource use.* Sinauer Associates, Sunderland, Mass.

Tracy, C. R. 1982. Biophysical modeling in reptilian physiology and ecology. In C. Gans and F. H. Pough, eds., *Biology of the Reptilia,* vol. 12, 275–321. Academic Press, New York.

———. 1991. Ecological responses of animals to climate. In R. Peters, ed., *Consequences of the greenhouse effect for biological diversity.* Yale University Press, New Haven, Conn.

Tracy, C. R., and T. L. George. 1992. On the determinants of extinction. *American Naturalist* 139:102–22.

Turner, F. B. 1970. The ecological efficiency of consumer populations. *Ecology* 51:741–42.

Turner, M. G. 1987. *Landscape heterogeneity and disturbance.* Springer-Verlag, New York.

Ulanowicz, R. E. 1989. Energy flow and productivity in the oceans. In P. J. Grubb and J. B. Whittaker, eds., *Toward a more exact ecology: The 30th symposium of the British Ecological Society, London 1988*, 327–51. Blackwell Scientific Publications, Oxford.

Valone, T. J., J. H. Brown, and E. J. Heske. 1993. Interactions between rodents and ants in the Chihuahuan Desert: An update. *Ecology* 75:252–55.

van den Bosch, F., R. Hengeveld, and J. A. J. Metz. 1992. Analyzing the velocity of animal range expansion. *Journal of Biogeography* 19:135–50.

Vandermeer, J. H., and R. H. MacArthur. 1966. A reformulation of alternative (b) of the broken stick model of species abundance. *Ecology* 47:139–40.

Van Riper, C. III, S. G. Van Riper, M. L. Goff, and M. Laird. 1986. The epizootiology and ecological significance of malaria in Hawaiian land birds. *Ecological Monographs* 56:327–35.

Van Valen, L. M. 1973a. Body size and numbers of plants and animals. *Evolution* 27:27–35.

———. 1973b. A new evolutionary law. *Evolutionary Theory* 1:1–30.

———. 1976. Energy and evolution. *Evolutionary Theory* 7:93–106.

Vermeij, G. J. 1978. *Biogeography and adaption: Patterns in marine life.* Harvard University Press, Cambridge, Mass.

———. 1987. *Evolution and escalation: An ecological history of life.* Princeton University Press, Princeton, N.J.

Vitousek, P. M., P. R. Ehrlich, A. H. Ehrlich, and P. A. Matson. 1986. Human appropriation of the products of photosynthesis. *BioScience* 36:368–73.

Vrba, E. S. 1983. Macroevolutionary trends: New perspectives on the roles of adaption and incidental effect. *Science* 221:387–89.

———. 1992. Mammals as a key to evolutionary theory. *Journal of Mammalogy* 73:1–28.

Vrba, E. S., and N. Eldredge. 1984. Individuals, hierarchies and processes: Towards a more complete evolutionary theory. *Paleobiology* 10:146–71.

Waldrup, M. M. 1992. *Complexity: The emerging science at the edge of order and chaos.* Simon & Schuster, New York.

Webb, S. D., and L. G. Marshall. 1982. Historical biogeography of recent South American land mammals. In M. A. Mares and H. H. Genoways, eds., *Mammalian biology in South America*, 39–52. Special Publication Series 6, Pymatuning Laboratory of Ecology, University of Pittsburgh.

Western, D., and M. Pearl, eds. 1989. *Conservation for the twenty-first century.* Oxford University Press, New York.

White, M. J. D. 1968. Models of speciation. *Science* 159:1065–70.

Whittaker, R. H. 1956. Vegetation of the Great Smoky Mountains. *Ecological Monographs* 22:1–44.

———. 1960. Vegetation of the Siskiyou Mountains, Oregon and California. *Ecological Monographs* 30:279–338.

———. 1967. Gradient analysis of vegetation. *Biological Reviews* 42:207–69.

———. 1970. *Communities and ecosystems.* Macmillan, New York.

Whittaker, R. H., and W. A. Niering. 1965. Vegetation of the Santa Catalina Mountains, Arizona: A gradient analysis of the south slope. *Ecology* 46:429–52.

Wiegert, R. In press. Energy and the dynamics of populations versus ecosystems. In C. G. Jones and J. H. Lawton, eds., *Linking species and ecosystems: Essays on the conjunction of perspectives.* Chapman and Hall, New York.

Wiens, J. A. 1973. Pattern and process in grassland bird communities. *Ecological Monographs* 43:237–70.

———. 1986. Spatial scale and temporal variation in studies of shrubsteppe birds. In J. Diamond and T. J. Case, eds., *Community ecology,* 154–72. Harper & Row, New York.

———. 1989. *The ecology of bird communities,* vols. I and II. Cambridge University Press, Cambridge.

Wiens, J. A., and M. I. Dyer. 1977. Assessing the potential impact of granivorous birds in ecosystems. In J. Pinowski and S. C. Kendeigh, eds., *Granivorous birds in ecosystems,* 205–66. Cambridge University Press, Cambridge.

Wiley, E. O. 1981. *Phylogenetics: The theory and practice of phylogenetic systematics.* Wiley, New York.

Wiley, E. O., and R. L. Mayden. 1985. Species and speciation in phylogenetic systematics, with examples from the North American fish fauna. *Annals of the Missouri Botanical Garden* 72:596–635.

Williams, C. B. 1964. *Patterns in the balance of nature and related problems in quantitative ecology.* Academic Press, New York.

Williams, G. C. 1992. *Natural selection: Domains, levels, and challenges.* Oxford University Press, Oxford.

Williamson, M. 1981. *Island populations.* Oxford University Press, Oxford.

Willis, J. C. 1922. *Age and area.* Cambridge University Press, Cambridge.

Wilson, D. S. 1975. The adequacy of body size as a niche difference. *American Naturalist* 109:769–84.

Wilson, D. S., and E. Sober. 1989. Reviving the superorganism. *Journal of Theoretical Biology* 136:356.

Wilson, E. O. 1985. The biological diversity crisis. *BioScience* 35:700–706.

———. 1988. The current state of biological diversity. In E. O. Wilson, ed., *Biodiversity,* 3–18. National Academy Press, Washington, D.C.

———. 1988. *Biodiversity.* National Academy Press, Washington, D.C.

Wright, D. H. 1983. Species-energy theory: An extension of species-area theory. *Oikos* 41:496–506.

———. 1987. Estimating human effects on global extinction. *International Journal of Biometeorology* 31:293–99.

———. 1990. Human impacts on energy flow through natural ecosystems, and implications for species endangerment. *Ambio* 19:189–94.

———. 1991. Correlations between incidence and abundance are expected by chance. *Journal of Biogeography* 18:463–66.

Zaret, T. M., and R. T. Paine. 1973. Species introduction in a tropical lake. *Science* 182:449–55.

Index